Principles and Practice of Engineering PE

CIVIL

Sample Questions & Solutions

Cover photography
Falls Park and Liberty Bridge in Greenville, South Carolina
Photographer: Patrick Wright
Designer: Pam Latour

Published by the
National Council of Examiners for Engineering and Surveying®
280 Seneca Creek Road, Clemson, SC 29631   800-250-3196   www.ncees.org

ISBN: 978-1-932613-31-5

Printed in the United States of America
November 2007

TABLE OF CONTENTS

NEWS AND UPDATES ON *www.ncees.org*

For news and updates about the examinations—including current exam specifications and design standards, exam policies, calculators approved for use during the examination, exam-day policies, scoring, errata for this book, and other information—visit the NCEES Web site at **www.ncees.org.**

NCEES—THE EXAM DEVELOPER

The Council:

- was established to assist and support its member licensing boards, which are located in each of the states as well as in the District of Columbia, Guam, the Northern Mariana Islands, Puerto Rico, and the Virgin Islands.

- **develops the examinations** required of candidates for licensure as professional engineers. These examinations measure a candidate's ability to demonstrate minimum competency in the practice of engineering and are administered by each NCEES member licensing board.

- follows the guidelines established in the *Standards for Educational and Psychological Testing* published by the American Psychological Association. These procedures maximize the fairness and quality of the examinations. To ensure that the procedures are followed, NCEES **uses experienced testing specialists** who have the expertise to guide the development of examinations using current testing techniques.

- **relies on committees composed of professional engineers** from throughout the nation to prepare the examinations. These licensed engineers—who come from diverse professional backgrounds including government, industry, private consulting, and academia—supply the content expertise that is essential in developing examinations.

LICENSING REQUIREMENTS

Licensure protects the public because it requires candidates to demonstrate certain qualifications before being allowed to practice as an engineering professional.

Eligibility

While examinations offer one means of measuring professional competency, most licensing boards also screen candidates based on education and experience. Because these requirements vary by state, we recommend that candidates contact the appropriate board. Board addresses and telephone numbers are listed on the NCEES Web site.

Application Procedures and Deadlines

Exam application procedures are available from the individual boards. Requirements and fees vary among the boards. Applicants are responsible for contacting their board office. Sufficient time must be allotted to complete the application process and assemble required data.

DESCRIPTION OF EXAMINATIONS

Exam Schedule

The Principles and Practice of Engineering (PE) examination in civil engineering is offered in the spring and fall of each year. The NCEES Web site lists exam dates for the next 10 years. You should contact your board for specific locations of exam sites.

Exam Content

The PE Civil exam is administered in two sessions, each lasting 4 hours. The morning session covers the breadth of civil engineering, and all examinees take the morning session. In the afternoon session examinees choose one of five civil modules: construction, geotechnical, structural, transportation, or water resources and environmental.

The examination consists of 80 equally weighted multiple-choice questions. Because there is no penalty for marking incorrect responses, candidates are advised to answer each question on the examination. Only one response should be marked for each question. No credit is given where two or more responses are marked.

This book presents half the number of questions included in an actual examination. Correct responses require reasoning and calculation demonstrating competent engineering judgment. Because they illustrate the general content covered in the examination, these questions should be helpful in preparing for the examination. Solutions are presented for all questions in this book, although the solution presented may not be the only way to solve a particular question. The intent is to demonstrate the typical effort required to solve each question. No representation is made or intended as to future exam questions, content, or subject matter.

Exam Development and Specifications

To be a valid measure of professional engineering competency, the PE examination must test knowledge pertaining to the specific tasks performed by professional engineers. NCEES ensures this by conducting regular surveys of licensed practitioners. The information gathered from these surveys:

- directs the exam development process
- is used to develop the exam content outline that is presented in the next section of this book
- determines the percentage of questions devoted to each subject area in the exam content outline.

Scoring Procedure

One of the most critical considerations in developing and administering examinations is establishing passing scores that reflect a standard of minimum competency. NCEES defines minimum competency as

> the minimum level of knowledge and skills a person must demonstrate in order to practice engineering and be in responsible charge in a manner that protects the health, safety, and welfare of the public.

Before setting a minimum passing score for a new exam or for the first exam after a change in the specifications or standards, NCEES conducts studies involving a representative panel of engineers familiar with the examinee population. This panel uses procedures widely recognized and accepted for occupational licensing purposes and develops a written standard of minimum competency that clearly articulates what skills and knowledge are required of licensed engineers. Panelists then take the examination, evaluating the difficulty level of each question in the context of the minimum competency standard. Finally, NCEES reviews the panel's work and sets the passing score for the initial exam. For subsequent exams, an equating method is used to set the passing score. The passing (raw) score is never disclosed.

NCEES does not use a fixed-percentage pass rate. The key issue is whether an individual candidate is competent to practice, **not** whether the candidate is better or worse than other candidates. To avoid the confusion that might arise from fluctuations in the passing score, exam results are reported simply as *pass* or *fail*. Some licensing jurisdictions may choose to report exam results of failing candidates as a scaled score.

The legal authority for making licensure decisions rests with the individual licensing boards and not with NCEES.

EXAM POLICIES AND PROCEDURES

A breach of an examination could lead to the licensure of people who are not competent to practice engineering. This puts the health, safety, and welfare of the public at risk. Therefore, NCEES takes measures necessary to protect the integrity of the exam process. This includes, for example, restricting cell phones, certain calculators, pencils, loose sheets of paper, and recording devices; controlling access into and out of the exam site; and monitoring activity in and around the exam room. Violating exam policies could result in such measures as dismissal from the exam, cancellation of exam results, and, in some cases, criminal action.

Be sure that you understand the policies outlined in the Candidate Agreement, and read all instructions from your board or testing service before exam day so that you know exactly what the expectations for examinees are.

Candidate Agreement

The Candidate Agreement explains the policies, procedures, and conditions examinees must agree to while taking an NCEES examination. Examinees are required to sign a statement on their answer sheet before the examination starts to affirm that they have been provided the

Candidate Agreement, have read and understand the material, and agree to abide by the conditions cited. These conditions apply to all NCEES examinations. A current Candidate Agreement is available on the NCEES Web site.

Special Accommodations

If you require special accommodations in the test-taking procedure, you should contact your state licensing board office well in advance of the day of the examination so that appropriate arrangements may be determined. Only preapproved accommodations are allowed on exam day.

Exam Admission Requirements

For exam admission, examinees must present a current, signed, government-issued photographic identification (such as a valid state driver's license or passport). Student IDs are not acceptable. Examinees must report to the exam site by the designated time. Examinees will not be admitted after the proctor begins reading the exam instructions.

Starting and Completing the Examination

Before the morning and afternoon sessions, proctors will distribute exam books containing an answer sheet. You should not open the exam book until instructed to do so by the proctor. Read the instructions and information given on the front and back covers and enter your name on the front cover. Listen carefully to all the instructions the proctor reads. The proctor has final authority on the administration of the examination.

The answer sheets for the multiple-choice questions are machine scored. For proper scoring, the answer spaces should be blackened completely using only the mechanical pencil provided to you. If you decide to change an answer, you must erase the first answer completely. Incomplete erasures and stray marks may be read as intended answers. One side of the answer sheet is used to collect identification and biographical data that may be used to analyze the performance of the examination.

The biographical data has no impact on the exam results. Proctors will guide you through the process of completing this portion of the answer sheet prior to taking the examination. This process will take approximately 15 minutes.

If you complete the examination with more than 15 minutes remaining, you are free to leave after returning all exam materials to the proctor. If you finish within 15 minutes of the end of the examination, you are required to remain until the end to avoid disruption to those still working and to permit orderly collection of all exam materials. Regardless of when you complete the examination, you are responsible for returning your assigned exam book. Examinees are not allowed to leave until the proctor has verified that all materials have been collected.

References

The PE examination is open-book. Examinees must bring their own reference materials, including design standards. All reference materials must be bound and remain bound during the examination. Individual licensing boards determine the reference materials allowed, so you should contact your board or testing service for specific advice.

Prohibited Items

A current list of prohibited items is included in the Candidate Agreement on the NCEES Web site. If a prohibited item is found in an examinee's possession after the exam begins, or if an examinee is found to be using a writing instrument other than the NCEES-supplied pencil, the item will be confiscated, the examinee will be dismissed from the exam, his or her exam will not be scored, and no refund will be provided. All confiscated items will be sent to NCEES.

Exam Irregularities

Fraud, deceit, dishonesty, and other irregular behavior in connection with taking any NCEES examination is strictly prohibited. Irregular behavior includes but is not limited to the following:

- Copying or allowing the copying of exam answers
- Failing to work independently
- Possessing unauthorized devices or source materials
- Surrogate testing or other dishonest conduct
- Disrupting other examinees
- Creating safety concerns
- Beginning the exam before the proctor instructs you to do so
- Failing to cease work on the examination or put down the pencil when time is called
- Possessing, reproducing, or disclosing exam questions, answers, or other information about the examination without authorization before, during, or after the exam administration
- Communicating with other examinees or with any outside source during the examination by telephone, personal computer, Internet, or any other means

Exam Results

Examinees are understandably eager to find out how they performed on the exam. To ensure that the process is fair and equitable to examinees and to maintain the validity of the exam questions, NCEES uses a rigorous scoring process for each of the NCEES multiple-choice examinations that takes approximately 12 weeks to complete.

- First, NCEES scans all answer sheets as they are received from the states. Answer sheets are flagged for review when they are missing critical information, such as the depth module examinees worked or the candidate ID number. The scoring process continues only when these issues are resolved.

- Next, a psychometric analysis is performed on a sample population of answer sheets from each multiple-choice examination to identify any questions with unusual statistics, which flag the question for review.

- Then, at least two subject-matter experts who are licensed engineers review the flagged items. In addition, NCEES reviews all Candidate Comment Forms, and the subject-matter experts consider comments on the forms about specific exam questions. If the reviews confirm an error in a question, credit may be given for more than one answer.

- When the analyses and reviews are completed, NCEES revises the answer keys as necessary. The passing score and the final correct answers for each exam are then used to score all the answer sheets. Scanners are calibrated before and during scoring. A percentage of the answer sheets are hand-graded and the results compared to the machine score to ensure accuracy of results.

- Finally, NCEES releases the results to the state boards or testing agencies, who in turn report the results to examinees.

PE CIVIL DESIGN STANDARDS AND EXAM SPECIFICATIONS

PE CIVIL DESIGN STANDARDS

The PE Civil exam refers to construction engineering, structural, and transportation design standards. A current list of applicable standards is shown on the NCEES Web site. Only printed, bound versions of the standards are allowed in the exam room. You will not be supplied any references during the exam, and you will not be allowed to exchange references with other examinees.

PE Civil BREADTH Exam Specifications
Effective Beginning with the April 2008 Examinations

		Approximate Percentage of Examination
I.	**CONSTRUCTION**	**20%**

A. Earthwork Construction and Layout
   1. Excavation and embankment (cut and fill)
   2. Borrow pit volumes
   3. Site layout and control

B. Estimating Quantities and Costs
   1. Quantity take-off methods
   2. Cost estimating

C. Scheduling
   1. Construction sequencing
   2. Resource scheduling
   3. Time-cost trade-off

D. Material Quality Control and Production
   1. Material testing (e.g., concrete, soil, asphalt)

E. Temporary Structures
   1. Construction loads

II.	**GEOTECHNICAL**	**20%**

A. Subsurface Exploration and Sampling
   1. Soil classification
   2. Boring log interpretation (e.g., soil profile)

B. Engineering Properties of Soils and Materials
   1. Permeability
   2. Pavement design criteria

C. Soil Mechanics Analysis
   1. Pressure distribution
   2. Lateral earth pressure
   3. Consolidation
   4. Compaction
   5. Effective and total stresses

D. Earth Structures
   1. Slope stability
   2. Slabs-on-grade

E. Shallow Foundations
   1. Bearing capacity
   2. Settlement

F. Earth Retaining Structures
   1. Gravity walls
   2. Cantilever walls
   3. Stability analysis
   4. Braced and anchored excavations

III. STRUCTURAL — 20%

A. Loadings
   1. Dead loads
   2. Live loads
   3. Construction loads

B. Analysis
   1. Determinate analysis

C. Mechanics of Materials
   1. Shear diagrams
   2. Moment diagrams
   3. Flexure
   4. Shear
   5. Tension
   6. Compression
   7. Combined stresses
   8. Deflection

D. Materials
   1. Concrete (plain, reinforced)
   2. Structural steel (structural, light gage, reinforcing)

E. Member Design
   1. Beams
   2. Slabs
   3. Footings

IV. TRANSPORTATION — 20%

A. Geometric Design
   1. Horizontal curves
   2. Vertical curves
   3. Sight distance
   4. Superelevation
   5. Vertical and/or horizontal clearances
   6. Acceleration and deceleration

V. WATER RESOURCES AND ENVIRONMENTAL **20%**

A. Hydraulics – Closed Conduit
  1. Energy and/or continuity equation (e.g., Bernoulli)
  2. Pressure conduit (e.g., single pipe, force mains)
  3. Closed pipe flow equations including Hazen-Williams, Darcy-Weisbach Equation
  4. Friction and/or minor losses
  5. Pipe network analysis (e.g., pipeline design, branch networks, loop networks)
  6. Pump application and analysis

B. Hydraulics – Open Channel
  1. Open-channel flow (e.g., Manning's equation)
  2. Culvert design
  3. Spillway capacity
  4. Energy dissipation (e.g., hydraulic jump, velocity control)
  5. Stormwater collection (e.g., stormwater inlets, gutter flow, street flow, storm sewer pipes)
  6. Flood plains/floodways
  7. Flow measurement – open channel

C. Hydrology
  1. Storm characterization (e.g., rainfall measurement and distribution)
  2. Storm frequency
  3. Hydrographs application
  4. Rainfall intensity, duration, and frequency (IDF) curves
  5. Time of concentration
  6. Runoff analysis including Rational and SCS methods
  7. Erosion
  8. Detention/retention ponds

D. Wastewater Treatment
  1. Collection systems (e.g., lift stations, sewer networks, infiltration, inflow)

E. Water Treatment
  1. Hydraulic loading
  2. Distribution systems

Total **100%**

Notes

1. The examination is developed with questions that will require a variety of approaches and methodologies including design, analysis, and application. Some questions may require knowledge of engineering economics.
2. The knowledge areas specified under 1, 2, 3, etc., are examples of kinds of knowledge, but they are not exclusive or exhaustive categories.
3. The breadth (AM) exam contains 40 multiple-choice questions. Examinee works all questions.
4. Score results are combined with depth exam results for final score.

PE Civil—CONSTRUCTION Depth Exam Specifications
Effective Beginning with the April 2008 Examinations

		Approximate Percentage of Examination
I.	**Earthwork Construction and Layout** A. Excavation and embankment (cut and fill) B. Borrow pit volumes C. Site layout and control D. Earthwork mass diagrams	**10%**
II.	**Estimating Quantities and Costs** A. Quantity take-off methods B. Cost estimating C. Engineering economics 1. Value engineering and costing	**17.5%**
III.	**Construction Operations and Methods** A. Lifting and rigging B. Crane selection, erection, and stability C. Dewatering and pumping D. Equipment production E. Productivity analysis and improvement F. Temporary erosion control	**15%**
IV.	**Scheduling** A. Construction sequencing B. CPM network analysis C. Activity time analysis D. Resource scheduling E. Time-cost trade-off	**17.5%**
V.	**Material Quality Control and Production** A. Material testing (e.g., concrete, soil, asphalt) B. Welding and bolting testing C. Quality control process (QA/QC) D. Concrete mix design	**10%**
VI.	**Temporary Structures** A. Construction loads B. Formwork C. Falsework and scaffolding D. Shoring and reshoring E. Concrete maturity and early strength evaluation F. Bracing G. Anchorage	**12.5%**

H. Cofferdams (systems for temporary excavation support)
I. Codes and standards [e.g., American Society of Civil Engineers (ASCE 37), American Concrete Institute (ACI 347), American Forest and Paper Association-NDS, Masonry Wall Bracing Standard]

VII. Worker Health, Safety, and Environment **7.5%**

A. OSHA regulations
B. Safety management
C. Safety statistics (e.g., incident rate, EMR)

VIII. Other Topics **10%**

A. Groundwater and well fields
   1. Groundwater control including drainage, construction dewatering
B. Subsurface exploration and sampling
   1. Drilling and sampling procedures
C. Earth retaining structures
   1. Mechanically stabilized earth wall
   2. Soil and rock anchors
D. Deep foundations
   1. Pile load test
   2. Pile installation
E. Loadings
   1. Wind loads
   2. Snow loads
   3. Load paths
F. Mechanics of materials
   1. Progressive collapse
G. Materials
   1. Concrete (prestressed, post-tensioned)
   2. Timber
H. Traffic safety
   1. Work zone safety

Total **100%**

Notes

1. The examination is developed with problems that will require a variety of approaches and methodologies including design, analysis, and application. Some problems may require knowledge of engineering economics.
2. The knowledge areas specified under A, B, C, etc., are examples of kinds of knowledge, but they are not exclusive or exhaustive categories.
3. Each depth (PM) exam contains 40 multiple-choice questions. Examinee chooses **one** depth examination and works all questions in the depth examination chosen.
4. Score results are combined with breadth exam results for final score.

PE Civil—GEOTECHNICAL Depth Exam Specifications

Effective Beginning with the April 2008 Examinations

		Approximate Percentage of Examination
I.	**Subsurface Exploration and Sampling**	**7.5%**
	A. Drilling and sampling procedures	
	B. Soil classification	
	C. General rock characterization (e.g., RQD, description, joints and fractures)	
	D. Boring log interpretation (e.g., soil profile)	
	E. In situ testing	
II.	**Engineering Properties of Soils and Materials**	**12.5%**
	A. Index properties	
	B. Phase relationships	
	C. Permeability	
	D. Geosynthetics	
	E. Pavement design criteria	
	F. Shear strength properties	
	G. Frost susceptibility	
III.	**Soil Mechanics Analysis**	**12.5%**
	A. Pressure distribution	
	B. Lateral earth pressure	
	C. Consolidation	
	D. Compaction	
	E. Expansive soils	
	F. Effective and total stresses	
	G. Seepage (e.g., exit gradient, drain fields, seepage forces, flow nets)	
IV.	**Earthquake Engineering**	**5%**
	A. Liquefaction	
	B. Pseudo-static analysis	
	C. Seismic site characterization	
V.	**Earth Structures**	**10%**
	A. Slope stability	
	B. Slabs-on-grade	
	C. Earth dams	
	D. Techniques and suitability of ground modification	

VI.	**Shallow Foundations**	**15%**
	A. Bearing capacity	
	B. Settlement	
	C. Mat and raft foundations	
VII.	**Earth Retaining Structures**	**15%**
	A. Gravity walls	
	B. Cantilever walls	
	C. Stability analysis	
	D. Mechanically stabilized earth walls	
	E. Braced and anchored excavations	
	F. Soil and rock anchors	
VIII.	**Deep Foundations**	**10%**
	A. Axial capacity (single pile/drilled shaft)	
	B. Lateral capacity and deflections (single pile/drilled shaft)	
	C. Settlement	
	D. Behavior of pile and/or drilled shaft group	
	E. Pile load test	
	F. Pile installation	
	G. Pile dynamics (e.g., wave equation, PDA test)	
IX.	**Other Topics**	**12.5%**
	A. Groundwater and well fields	
	1. Well logging and subsurface properties	
	2. Aquifers (e.g., characterization)	
	3. Groundwater flow including Darcy's Law and seepage analysis	
	4. Well analysis (steady flow only)	
	5. Groundwater control including drainage, construction dewatering	
	B. Loadings	
	1. Earthquake loads	
	C. Construction operations and methods	
	1. Dewatering and pumping	
	2. Quality control process (QA/QC) (e.g., when digging, confirming quality; writing QA processes)	
	D. Temporary structures	
	1. Shoring and reshoring	
	2. Concrete maturity and early strength evaluation	
	3. Bracing	
	4. Anchorage	
	5. Cofferdams (systems for temporary excavation support)	
	E. Worker health, safety, and environment	
	1. OSHA regulations	
	2. Safety management	
Total		**100%**

Notes

1. The examination is developed with problems that will require a variety of approaches and methodologies including design, analysis, and application. Some problems may require knowledge of engineering economics.
2. The knowledge areas specified under 1, 2, 3, etc., are examples of kinds of knowledge, but they are not exclusive or exhaustive categories.
3. Each depth (PM) exam contains 40 multiple-choice questions. Examinee chooses **one** depth examination and works all questions in the depth examination chosen.
4. Score results are combined with breadth exam results for final score.

PE Civil—STRUCTURAL Depth Exam Specifications

Effective Beginning with the April 2008 Examinations

		Approximate Percentage of Examination
I.	**Loadings**	**12.5%**
	A. Dead loads	
	B. Live loads	
	C. Construction loads	
	D. Wind loads	
	E. Earthquake loads, including liquefaction, site characterization, and pseudo-static analysis	
	F. Moving loads	
	G. Snow loads	
	H. Impact loads	
	I. Load paths	
	J. Load combinations	
II.	**Analysis**	**12.5%**
	A. Determinate analysis	
	B. Indeterminate analysis	
III.	**Mechanics of Materials**	**12.5%**
	A. Shear diagrams	
	B. Moment diagrams	
	C. Flexure	
	D. Shear	
	E. Tension	
	F. Compression	
	G. Combined stresses	
	H. Deflection	
	I. Progressive collapse	
	J. Torsion	
	K. Buckling	
	L. Fatigue	
	M. Thermal deformation	
IV.	**Materials**	**12.5%**
	A. Concrete (plain, reinforced)	
	B. Concrete (prestressed, post-tension)	
	C. Structural steel (structural, light gage, reinforcing)	
	D. Timber	
	E. Masonry (brick veneer, CMU)	
	F. Composite construction	

V. Member Design **25%**

A. Beams
B. Slabs
C. Footings
D. Columns
E. Trusses
F. Braces
G. Frames
H. Connections (bolted, welded, embedded, anchored)
I. Shear walls
J. Diaphragms (horizontal, vertical, flexible, rigid)
K. Bearing walls

VI. Design Criteria **12.5%**

A. International Building Code (IBC)
B. American Concrete Institute (ACI-318, 530)
C. Precast/Prestressed Concrete Institute (PCI Design Handbook)
D. Manual of Steel Construction (AISC) including AISC 341
E. National Design Specification for Wood Construction (NDS)
F. Standard Specifications for Highway Bridges (AASHTO)
G. American Society of Civil Engineers (ASCE-7)
H. American Welding Society (AWS D1.1, D1.2, and D1.4)

VII. Other Topics **12.5%**

A. Engineering properties of soils and materials
    1. Index properties (e.g., plasticity index; interpretation and how to use them)
B. Soil mechanics analysis
    1. Expansive soils
C. Shallow foundations
    1. Mat and raft foundations
D. Deep foundations
    1. Axial capacity (single pile/drilled shaft)
    2. Lateral capacity and deflections (single pile/drilled shaft)
    3. Settlement
    4. Behavior of pile and/or drilled shaft group
E. Engineering Economics
    1. Value engineering and costing
F. Material Quality Control and Production
    1. Welding and bolting testing
G. Temporary Structures
    1. Formwork
    2. Falsework and scaffolding
    3. Shoring and reshoring
    4. Concrete maturity and early strength evaluation
    5. Bracing
    6. Anchorage

H. Worker Health, Safety and Environment
   1. OSHA regulations
   2. Safety management

Total **100%**

Notes

1. The examination is developed with problems that will require a variety of approaches and methodologies including design, analysis, and application. Some problems may require knowledge of engineering economics.
2. The knowledge areas specified under 1, 2, 3, etc., are examples of kinds of knowledge, but they are not exclusive or exhaustive categories.
3. Each depth (PM) exam contains 40 multiple-choice questions. Examinee chooses **one** depth exam and works all questions in the depth exam chosen.
4. Score results are combined with breadth exam results for final score.

PE Civil—TRANSPORTATION Depth Exam Specifications
Effective Beginning with the April 2008 Examinations

		Approximate Percentage of Examination
I.	**Traffic Analysis**	**22.5%**
	A. Traffic capacity studies	
	B. Traffic signals	
	C. Speed studies	
	D. Intersection analysis	
	E. Traffic volume studies	
	F. Sight distance evaluation	
	G. Traffic control devices	
	H. Pedestrian facilities	
	I. Driver behavior and/or performance	
II.	**Geometric Design**	**30%**
	A. Horizontal curves	
	B. Vertical curves	
	C. Sight distance	
	D. Superelevation	
	E. Vertical and/or horizontal clearances	
	F. Acceleration and deceleration	
	G. Intersections and/or interchanges	
III.	**Transportation Planning**	**7.5%**
	A. Optimization and/or cost analysis (e.g., transportation route A or transportation route B)	
	B. Traffic impact studies	
	C. Capacity analysis (future conditions)	
IV.	**Traffic Safety**	**15%**
	A. Roadside clearance analysis	
	B. Conflict analysis	
	C. Work zone safety	
	D. Accident analysis	
V.	**Other Topics**	**25%**
	A. Hydraulics	
	1. Flow measurement – closed conduits	
	2. Open channel – subcritical and supercritical flow	
	B. Hydrology	
	1. Hydrograph development and synthetic hydrographs	
	C. Engineering properties of soils and materials	
	1. Index properties (e.g., identification of types of soils; suitable or unsuitable)	

D. Soil mechanics analysis
    1. Expansive soils
E. Engineering economics
    1. Value engineering and costing
F. Construction operations and methods
    1. National Pollutant Discharge Elimination System (NPDES) permitting
G. Temporary structures
    1. Concrete maturity and early strength evaluation

Notes

1. The examination is developed with problems that will require a variety of approaches and methodologies including design, analysis, and application. Some problems may require knowledge of engineering economics.
2. The knowledge areas specified as examples of kinds of knowledge are not exclusive or exhaustive categories.
3. Each depth (PM) exam contains 40 multiple-choice questions. Examinee chooses **one** depth exam and works all questions in the depth exam chosen.
4. Score results are combined with breadth exam results for final score.

PE Civil—WATER RESOURCES and ENVIRONMENTAL Depth Exam Specifications

Effective Beginning with the April 2008 Examinations

		Approximate Percentage of Examination
I.	**Hydraulics – Closed Conduit**	**15%**
	A. Energy and/or continuity equation (e.g., Bernoulli)	
	B. Pressure conduit (e.g., single pipe, force mains)	
	C. Closed pipe flow equations including Hazen-Williams, Darcy-Weisbach Equation	
	D. Friction and/or minor losses	
	E. Pipe network analysis (e.g., pipeline design, branch networks, loop networks)	
	F. Pump application and analysis	
	G. Cavitation	
	H. Transient analysis (e.g., water hammer)	
	I. Flow measurement – closed conduits	
	J. Momentum equation (e.g., thrust blocks, pipeline restraints)	
II.	**Hydraulics – Open Channel**	**15%**
	A. Open-channel flow (e.g., Manning's equation)	
	B. Culvert design	
	C. Spillway capacity	
	D. Energy dissipation (e.g., hydraulic jump, velocity control)	
	E. Stormwater collection including stormwater inlets, gutter flow, street flow, storm sewer pipes	
	F. Flood plain/floodway	
	G. Subcritical and supercritical flow	
	H. Flow measurement – open channel	
	I. Gradually varied flow	
III.	**Hydrology**	**15%**
	A. Storm characterization including rainfall measurement and distribution	
	B. Storm frequency	
	C. Hydrographs application	
	D. Hydrograph development and synthetic hydrographs	
	E. Rainfall intensity, duration, and frequency (IDF) curves	
	F. Time of concentration	
	G. Runoff analysis including Rational and SCS methods	
	H. Gauging stations including runoff frequency analysis and flow calculations	
	I. Depletions (e.g., transpiration, evaporation, infiltration)	
	J. Sedimentation	
	K. Erosion	
	L. Detention/retention ponds	

IV. Groundwater and Well Fields **7.5%**

A. Aquifers (e.g., characterization)
B. Groundwater flow including Darcy's Law and seepage analysis
C. Well analysis (steady flow only)
D. Groundwater control including drainage, construction dewatering
E. Water quality analysis
F. Groundwater contamination

V. Wastewater Treatment **15%**

A. Wastewater flow rates (e.g., municipal, industrial, commercial)
B. Unit operations and processes
C. Primary treatment (e.g., bar screens, clarification)
D. Secondary clarification
E. Chemical treatment
F. Collection systems (e.g., lift stations, sewer network, infiltration, inflow)
G. National Pollutant Discharge Elimination System (NPDES) permitting
H. Effluent limits
I. Biological treatment
J. Physical treatment
K. Solids handling (e.g., thickening, drying processes)
L. Digesters
M. Disinfection
N. Nitrification and/or denitrification
O. Operations (e.g., odor control, corrosion control, compliance)
P. Advanced treatment (e.g., nutrient removal, filtration, wetlands)
Q. Beneficial reuse (e.g., liquids, biosolids, gas)

VI. Water Quality **15%**

A. Stream degradation (e.g., thermal, base flow, TDS, TSS, BOD, COD)
B. Oxygen dynamics (e.g., oxygenation, deoxygenation, oxygen sag curve)
C. Risk assessment and management
D. Toxicity
E. Biological contaminants (e.g., algae, mussels)
F. Chemical contaminants (e.g., organics, heavy metals)
G. Bioaccumulation
H. Eutrophication
I. Indicator organisms and testing
J. Sampling and monitoring (e.g., QA/QC, laboratory procedures)

VII. Water Treatment **15%**

A. Demands
B. Hydraulic loading
C. Storage (raw and treated water)
D. Sedimentation
E. Taste and odor control
F. Rapid mixing
G. Coagulation and flocculation

H. Filtration
I. Disinfection
J. Softening
K. Advanced treatment (e.g., membranes, activated carbon, desalination)
L. Distribution systems

VIII. Engineering Economics **2.5%**

A. Life-cycle modeling
B. Value engineering and costing

Notes

1. The examination is developed with problems that will require a variety of approaches and methodologies including design, analysis, and application. Some problems may require knowledge of engineering economics.
2. The knowledge areas specified under A, B, C, etc., are examples of kinds of knowledge, but they are not exclusive or exhaustive categories.
3. Each depth (PM) exam contains 40 multiple-choice questions. Examinee chooses **one** depth exam and works all questions in the depth exam chosen.
4. Score results are combined with breadth exam results for final score.

CIVIL BREADTH
MORNING SAMPLE QUESTIONS

This book contains 20 civil breadth questions, half the number on the actual exam.

MORNING SAMPLE QUESTIONS

101. Given the figure below, the net excavation (yd^3) from Station 1+00 to Station 3+00 is most nearly:

(A) 160
(B) 262
(C) 390
(D) 463

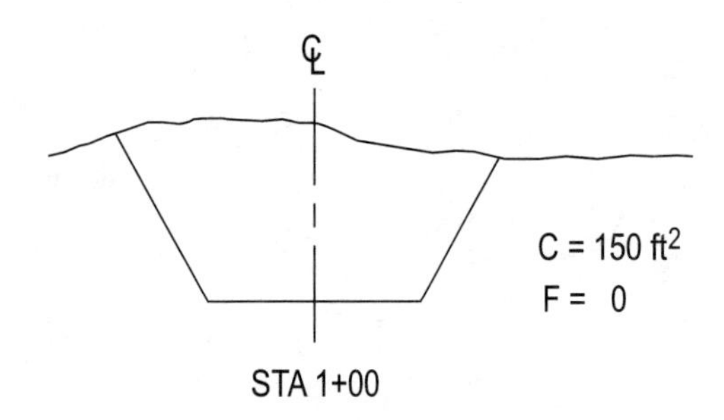

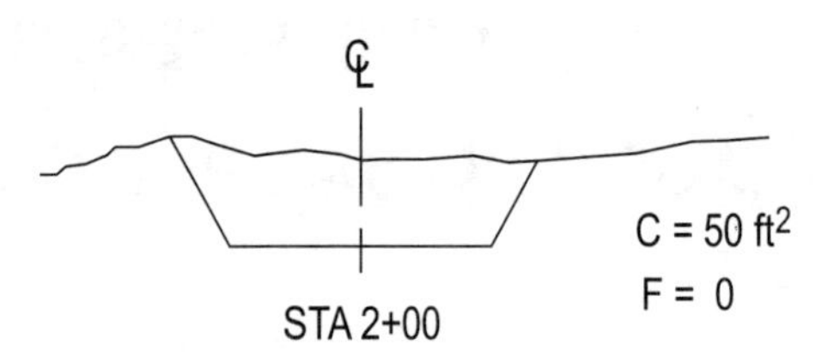

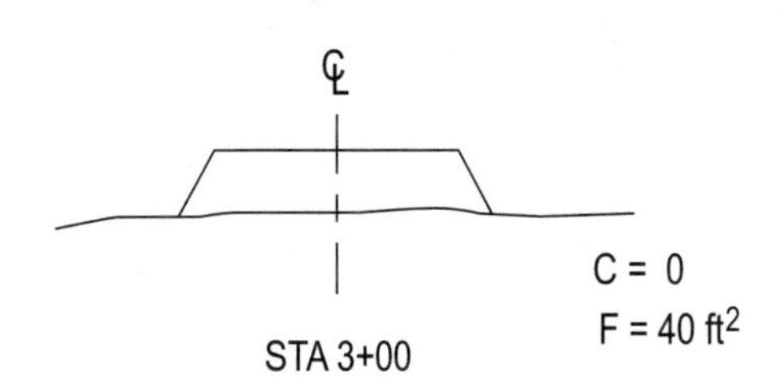

102. A 227-ft length of canal is to be lined with concrete for erosion control. With 12% waste and overbreak, the volume (yd^3) of concrete that must be delivered is most nearly:

(A) 234
(B) 280
(C) 292
(D) 327

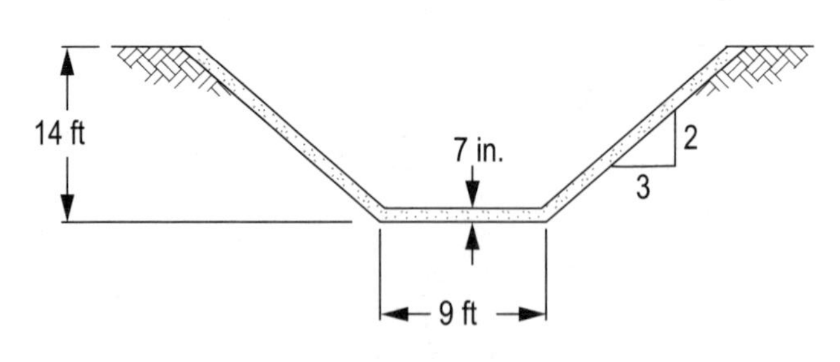

103. An activity on arrow network is shown below. Define variables as follows:

TF = Total float
LF = Late finish
LS = Late start
ES = Early start
EF = Early finish
D = Duration

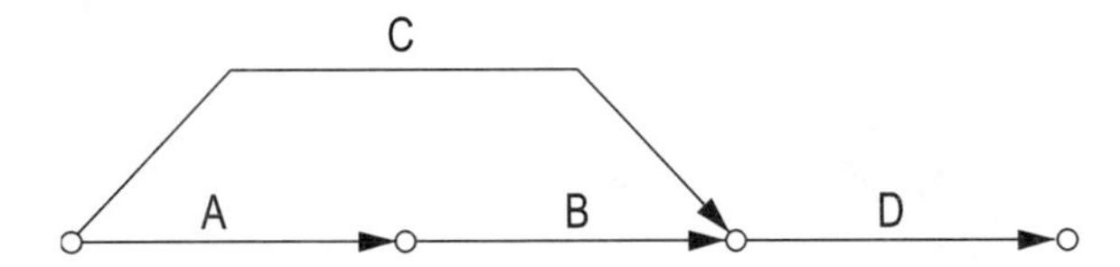

Using subscripts to denote the different activities, the expression for the total float of Activity B, TF_B, is most nearly:

(A) $LS_D - ES_B - D_B$
(B) $LS_B - EF_A$
(C) $ES_D - ES_B - D_B$
(D) $LF_B - ES_B - D_B$

104. If the water-to-cement ratio of concrete is decreased, which statement about the concrete is **NOT** true?

(A) Water tightness is decreased.
(B) Workability is decreased.
(C) Strength is increased.
(D) Durability is increased.

105. The effective overburden pressure (psf) at the middle of the clay layer shown in the subsurface section below is most nearly:

(A) 1,000
(B) 1,080
(C) 1,390
(D) 1,700

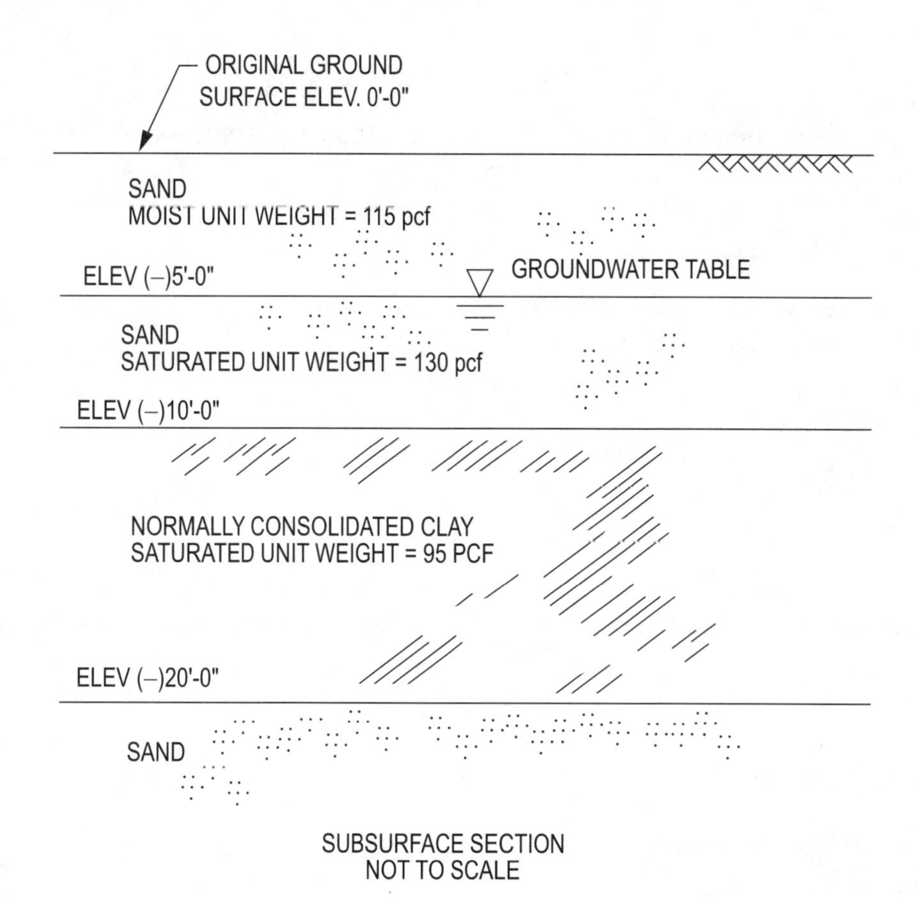

GO ON TO THE NEXT PAGE

106. Refer to the following test data:

% passing #200 sieve	60
Liquid limit	55
Plastic limit	30

The USCS Classification of this soil is:

(A) A-7-6
(B) clay loam
(C) CL
(D) CH

107. A soil has a Modified Proctor optimum moisture content of 14.0% and a Modified Proctor maximum dry density of 116.0 pcf. Assume that the soil is transported from a borrow pit to a construction site to construct 500,000 yd^3 of compacted roadway embankment. The soil arrives at the construction site with a moisture content of 9%. The soil is placed and compacted to 90% of the Modified Proctor maximum dry density. The total volume of water (gal) that must be added to the soil to increase the moisture content to the optimum level is most nearly:

(A) 313,000
(B) 4,500,000
(C) 8,500,000
(D) 9,400,000

108. A retaining wall will be constructed at a site. Assume the sand backfill has a friction angle of 28°. According to Rankine's theory, the active lateral earth pressure coefficient is most nearly:

(A) 0.30
(B) 0.35
(C) 0.40
(D) 0.50

GO ON TO THE NEXT PAGE

109. For the structure shown below, what load combination should be used to design the structure to resist the overturning and uplift of Footing A?

(A) Dead only
(B) Dead plus live
(C) Dead plus wind
(D) Dead plus live plus wind

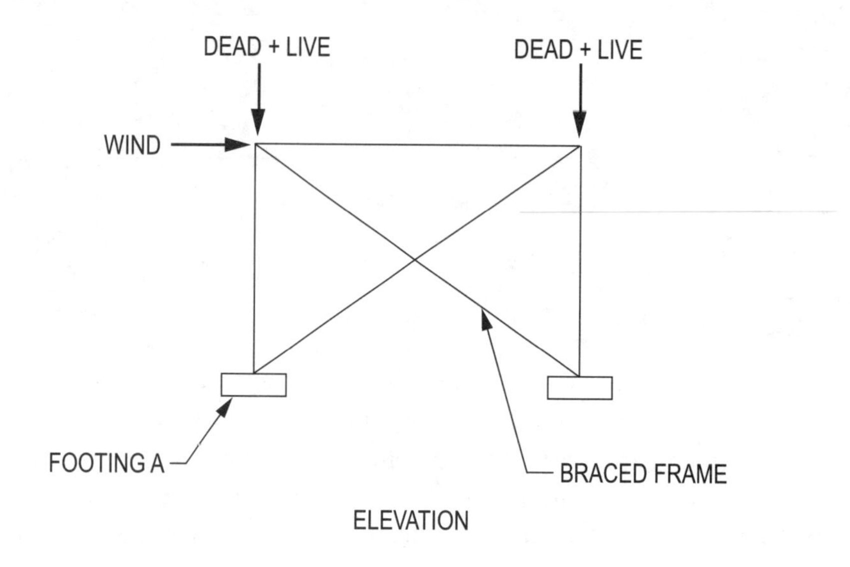

110. The concrete staircase shown in the figure below is proposed for an office building extension. The vertical reaction (kips) at A is most nearly:

(A) 7.0
(B) 8.0
(C) 8.5
(D) 9.8

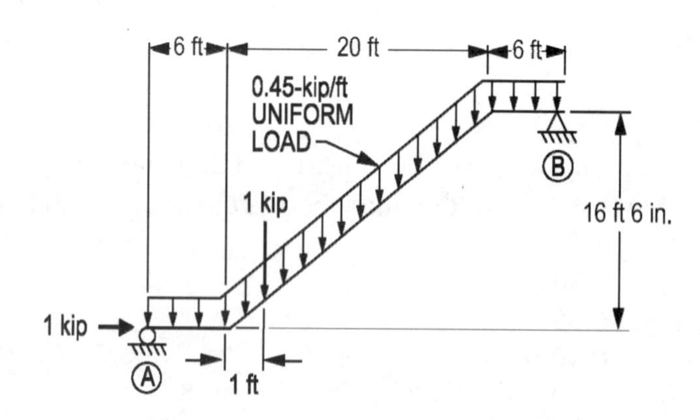

111. What would be the effect of changing the yield strength, F_y, of a steel beam from 50 ksi to 36 ksi while keeping the remaining design data the same?

(A) Deflection will increase.

(B) Deflection will decrease.

(C) Not enough information is given.

(D) There will be no change in deflections.

112. Assuming all steel areas are equal, which of the following steel sections is most efficient for use as a beam spanning 20 ft with an unbraced length of the compression flange of 20 ft?

(A) 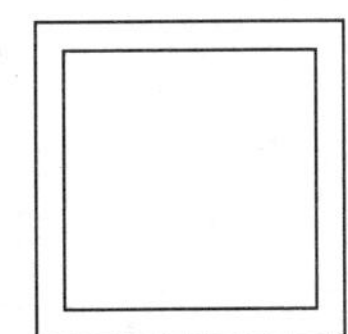

(B) 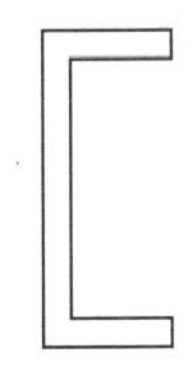

(C) 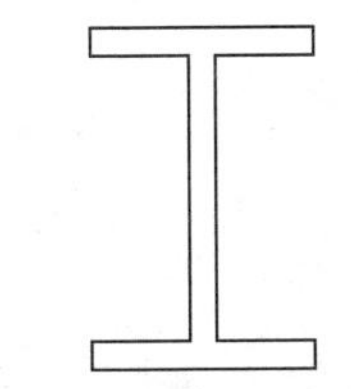

(D) 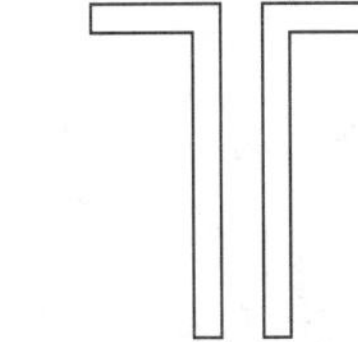

GO ON TO THE NEXT PAGE

113. The tangent vertical alignment of a section of proposed highway is shown in the figure below. The vertical clearance (ft) between the bridge structure at Station 73+00 and the vertical curve is most nearly:

(A) 15.3
(B) 19.0
(C) 19.8
(D) 22.1

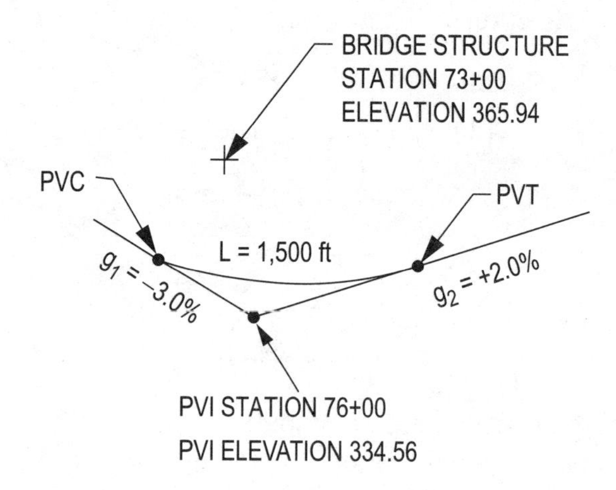

114. Level of service (LOS) is a quality measure describing operational conditions within a traffic stream, generally in terms of such service measures as speed travel time, freedom to maneuver, traffic interruptions, and comfort and convenience. How many levels of service are defined?

(A) Three

(B) Four

(C) Five

(D) Six

GO ON TO THE NEXT PAGE

115. Referring to the horizontal curve shown in the figure below, the station of the PI is most nearly:

(A) 29+94.90
(B) 33+42.84
(C) 34+77.82
(D) 35+10.47

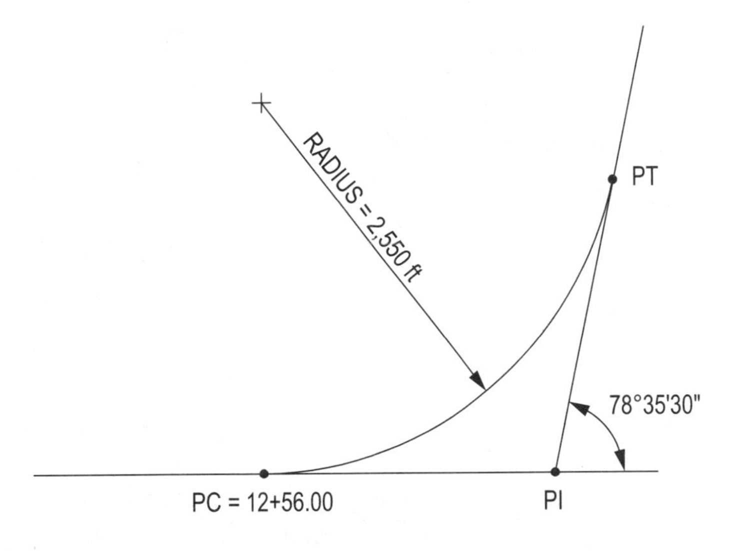

116. The tangent vertical alignment of a section of proposed highway is shown in the figure below. The horizontal distance (ft) between PVI_1 and PVI_2 is most nearly:

(A) 102
(B) 2,900
(C) 3,400
(D) 3,502

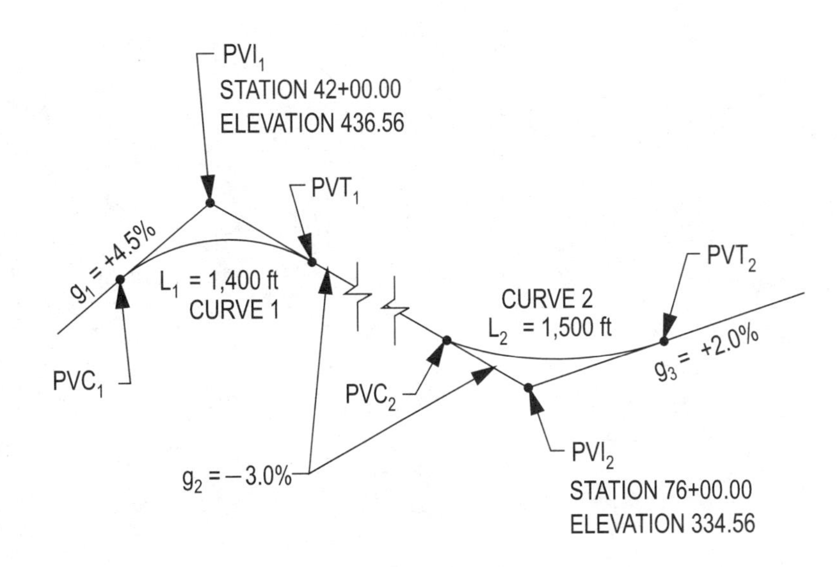

117. Given a net total dynamic head of 1,200 ft and a flow of 10 cfs, which type of single-stage impeller would be the most efficient to use?

(A) Radial flow

(B) Mixed flow

(C) Axial flow

(D) Positive displacement

118. A hydraulic jump is most likely to occur under which of the following conditions?

(A) The slope of an open channel with a uniform cross section and roughness coefficient suddenly changes from a high (steep) value to a low (mild) value. The flow rate (cfs) is constant.

(B) The roughness coefficient of an open channel with a uniform cross section and slope suddenly changes from a high value to a low value. The flow rate (cfs) is constant.

(C) The diameter of a pressure pipe system with a uniform roughness coefficient suddenly changes from 12 in. to 15 in. The flow rate (cfs) is constant.

(D) Flow is removed from an open channel with a uniform slope, roughness coefficient, and cross section.

GO ON TO THE NEXT PAGE

119. The 1-hour unit hydrograph for a watershed is given in the figure below. A 2-hour storm of intensity 0.5 in./hr in this watershed produces a total excess volume of water (acre-ft) of most nearly:

(A) 0.01
(B) 37
(C) 75
(D) 450

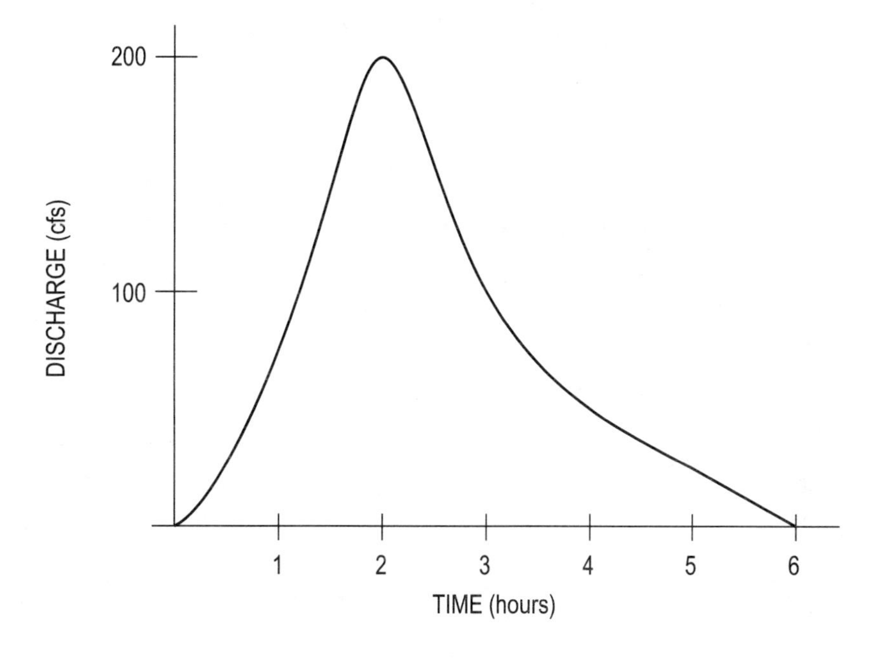

TIME (hours)	0	1	2	3	4	5	6
DISCHARGE (cfs)	0	75	200	100	50	25	0

120. A gravity sewer system is being designed for a minimum velocity of 2 ft/sec. What is the reason behind this practice?

(A) To prevent deposition of solids

(B) To release trapped sewer gases

(C) To ventilate the waste with turbulence

(D) To reduce the length of piping necessary

CONSTRUCTION
AFTERNOON SAMPLE QUESTIONS

This book contains 20 construction depth questions, half the number on the actual exam.

501. A segment of interstate highway requires the construction of an embankment of 500,000 yd^3. The embankment fill is to be compacted to a minimum of 90% of Modified Proctor maximum dry density.

A source of suitable borrow has been located for construction of the embankment. Assume that there is no soil loss in transporting the soil from the borrow pit to the embankment.

The following data apply:

Dry unit weight of soil in borrow pit	113.0 pcf
Moisture content in borrow pit	16.0%
Specific gravity of the soil particles	2.65
Modified Proctor optimum moisture content	13.0%
Modified Proctor maximum dry density	120.0 pcf

Assuming each truck holds 5.0 yd^3 and the void ratio of the soil is 1.30 during transport, the minimum number of truckloads of soil from the borrow pit that is required to construct the embankment is most nearly:

(A) 100,000
(B) 150,000
(C) 200,000
(D) 250,000

502. In the figure below, the radius R is 200 ft, and the mid-ordinate M is 12.8 ft. The length (ft) of the curve is most nearly:

(A) 143.9
(B) 140.8
(C) 75.2
(D) 71.9

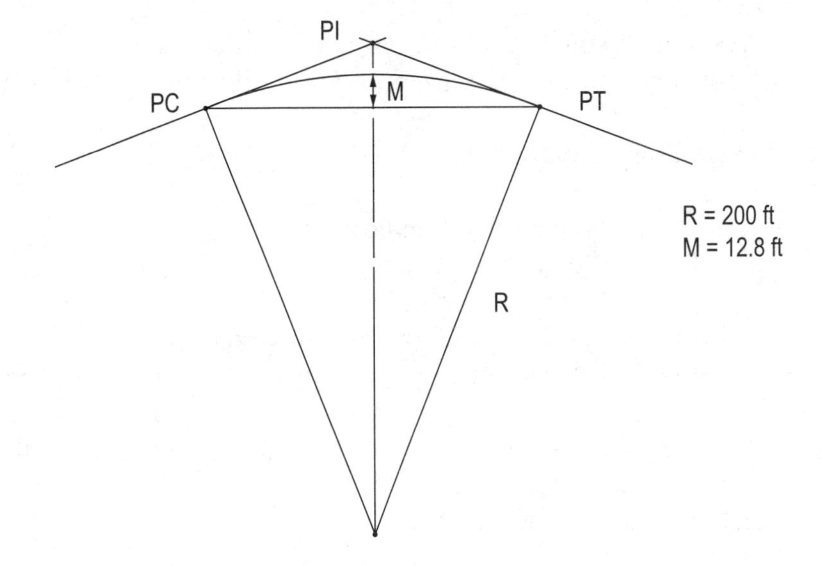

503. A 200-ft × 100-ft room has been prepared for painting. The walls are 7 ft high and will require two coats of paint on the previously painted surface for proper coverage. If 1 gal of paint covers 300 ft^2, the number of gallons of paint needed to paint the walls is most nearly:

(A) 14
(B) 28
(C) 67
(D) 81

GO ON TO THE NEXT PAGE

504. After purchasing a quarry and basic crushing equipment, the contractor is considering an alternative plan to improve the operation of the quarry. The alternative plan will produce an equal amount of crushed rock and equal revenue.

Parameter	Present Plan	Alternative Plan
First Cost ($)	0	10,000
Salvage ($)	0	1,000
Annual Cost ($)	250,000	248,000
Life (years)	—	5

The benefit-cost ratio of the alternative plan (using a 10% rate of return on investment) when compared to the present plan is most nearly:

(A) 0.6
(B) 0.8
(C) 1.0
(D) 1.2

GO ON TO THE NEXT PAGE

505. A free-standing concrete wall that is 72 ft long × 12 ft high × 1 ft thick is to be built in three equal pours. The following data apply.

Exclude overhead and profit.
Exclude wall bulkheads.
Material prices include all taxes and delivery.
Include 10% waste on concrete.

Labor

Carpenter	\$32.73/hr
Laborer	\$26.08/hr
Supervisor	\$35.37/hr

Productivity

Erect forms	5.5 ft^2/LH*
Strip forms	15.0 ft^2/LH
Place concrete	2.2 yd^3/LH

*LH = labor hour

Crews

Erect and strip	4 carpenters
	2 laborers
	1 supervisor (working)
Place concrete	3 laborers
	1 carpenter
	1 supervisor (working)

Materials

Formwork, initial erection	\$2.66/ft^2
reuse	\$0.34/ft^2
Concrete, Redi-Mix at site	\$97.20/yd^3
Reinforcing subcontract	\$120.00/yd^3

The total cost of the wall is most nearly:

(A) \$15,331
(B) \$19,153
(C) \$22,993
(D) \$25,665

506. The rigging shown below will be used to lift a 60-kip load. The center of gravity of the load is 40 ft from the left end. The force (kips) in Sling A is most nearly:

(A) 30
(B) 32.3
(C) 40
(D) 43.1

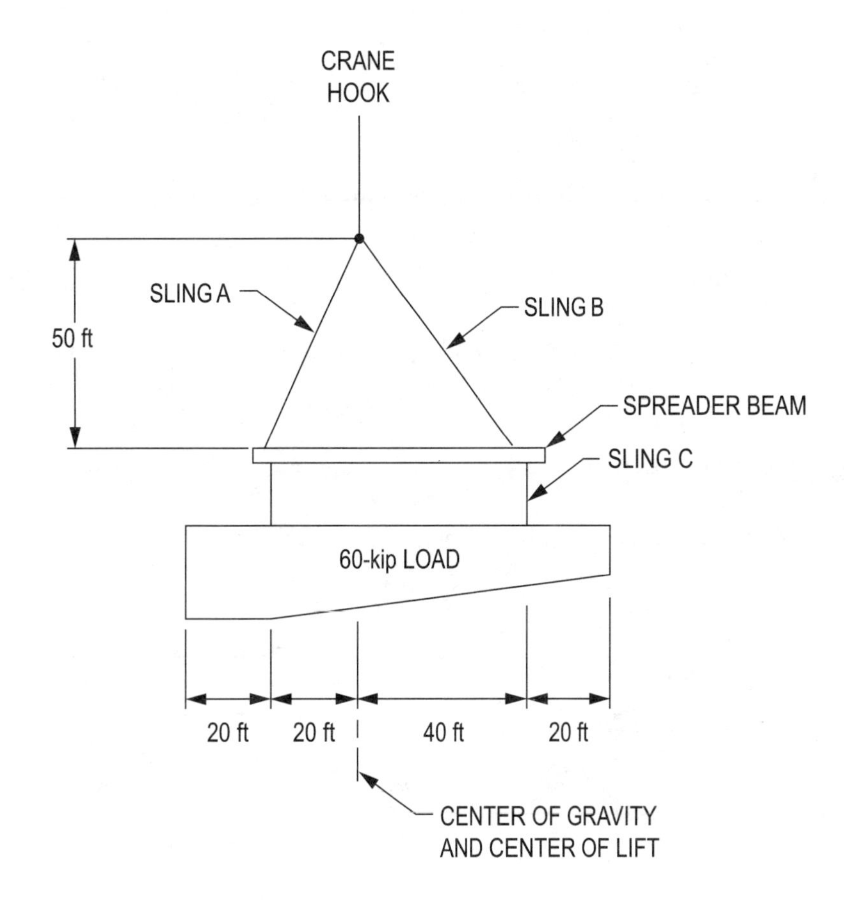

GO ON TO THE NEXT PAGE

507. A truck is hauling earth from a construction site. The truck has the following specifications:

Maximum allowable gross vehicle weight	37,800 lb
Empty vehicle weight	10,800 lb
Heaped capacity	12 yd^3
Struck capacity	10 yd^3
Haul time to dump area, including load, haul, return, and dump times	17 min
Delay time	5 min/hr

The soil has the following characteristics:

Bank density	110 pcf
Loose density	100 pcf

The productivity, in bank measure (yd^3/hr), of this operation is most nearly:

(A) 38.8
(B) 35.3
(C) 32.3
(D) 29.5

508. An activity-on-node network for a project is shown in the following figure. All relationships are finish-to-start with no lag unless otherwise noted. If all activities begin at their early start except Activity E, which is delayed by 2 days from its early start, which of the following statements is true?

(A) Activity E will have no impact on the start time of any other activity.

(B) Activity E will delay the start of Activity G by 1 day but will not delay project completion.

(C) Activity E will delay the start of Activity G by 2 days but will not delay project completion.

(D) Activity E will delay the completion of the project by 2 days.

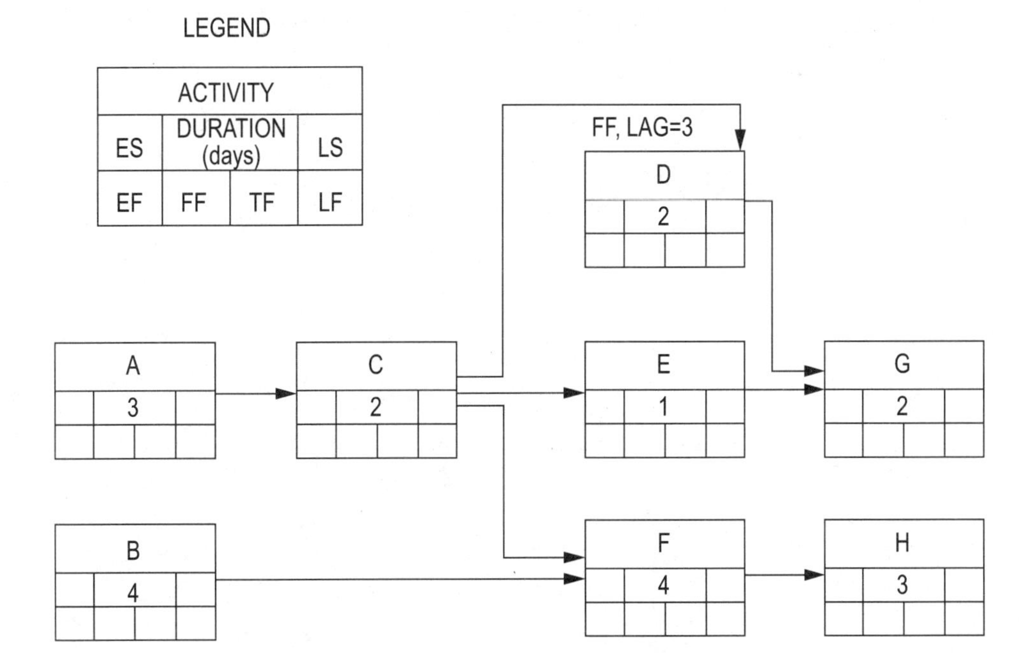

509. In the activity-on-arrow network below, the early start of Activity N is most nearly:

(A) 16
(B) 21
(C) 23
(D) 24

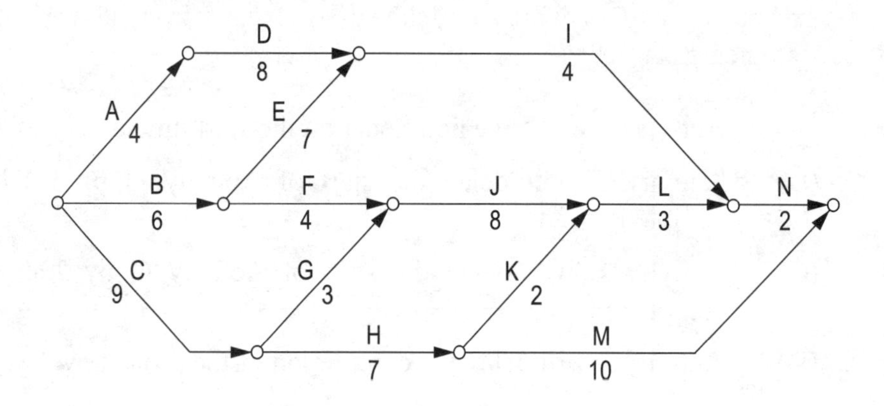

510. Referring to the grade profile and mass diagram for a roadway construction project shown below, which of the following is/are true?

I. The job is balanced (i.e., equal cut and fill).
II. Section B–D represents a fill operation.
III. Station D represents a transition point between cut and fill.

(A) I only
(B) II only
(C) III only
(D) II and III only

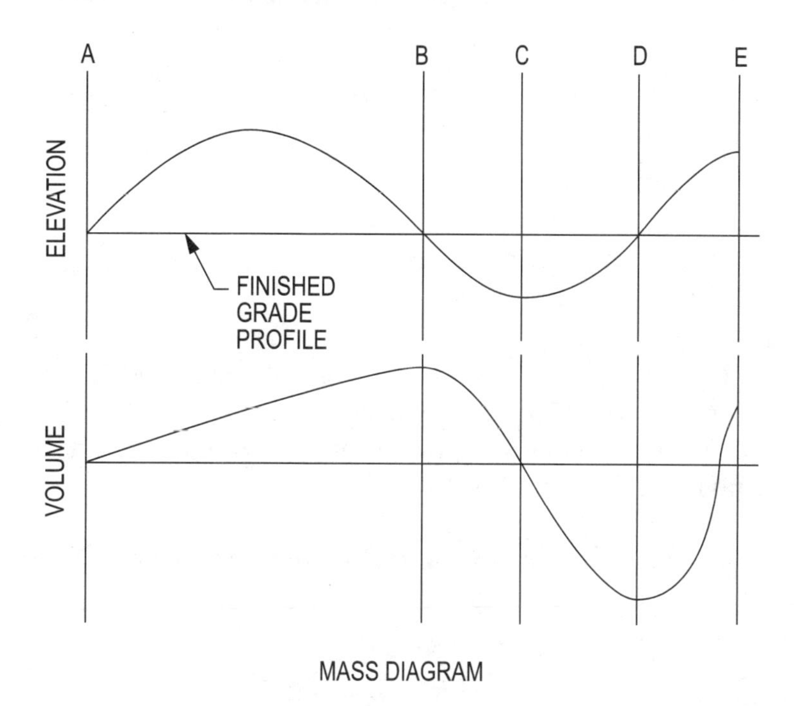

511. A formal CPM analysis for a project shows the planned costs to date are $85,000, and the accounting department reports charges to the job of $90,000. If the reported earned value to date is $70,000, the cost and schedule status of the project are most nearly:

(A) ahead of schedule and over budget

(B) behind schedule and over budget

(C) ahead of schedule and under budget

(D) behind schedule and under budget

512. A highway project requires a concrete mix. On a weight basis, the design mix has the proportions 1:2.25:3.25. Cement content was specified at 6.28 sacks/yd^3. The aggregates are SSD and have specific gravities of 2.65 for both the fine and the coarse aggregate. The specific gravity of the cement is 3.15.

The water/cement ratio (gal/sack) of the concrete mix is most nearly:

(A) 3.6
(B) 4.2
(C) 4.7
(D) 5.2

513. The figure shows the basement section of a building. The water table is above the footings. The walls must support earth for lateral earth pressure. A triangular pressure distribution will be used for the design. The following equivalent fluid densities apply:

Pressure	Above Water Table	Below Water Table
Active	27 pcf	21 pcf
At rest	45 pcf	35 pcf
Passive	300 pcf	230 pcf

Assume the soil above the water table is saturated.

Considering only one of the basement walls and only that portion between the floor and the top of the footing, the total applied lateral force (lb/ft) is most nearly:

(A) 3,800
(B) 4,800
(C) 5,300
(D) 5,900

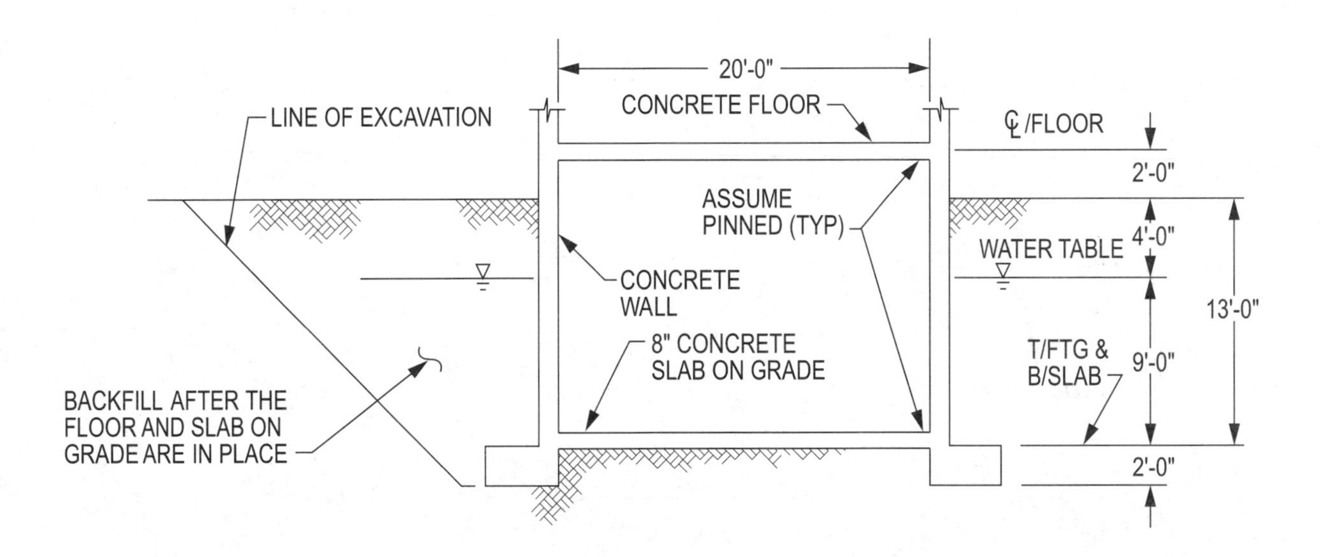

514. An 8-ft-high concrete retaining wall, battered on one side at a ratio of 1:12, will be poured with 150-pcf concrete to its full height in less than 1 hour. The calculated maximum concrete pressure in the form is 1,200 psf. The uplift force (lb/ft) the form will be subjected to is most nearly:

(A) 0
(B) 33
(C) 400
(D) 800

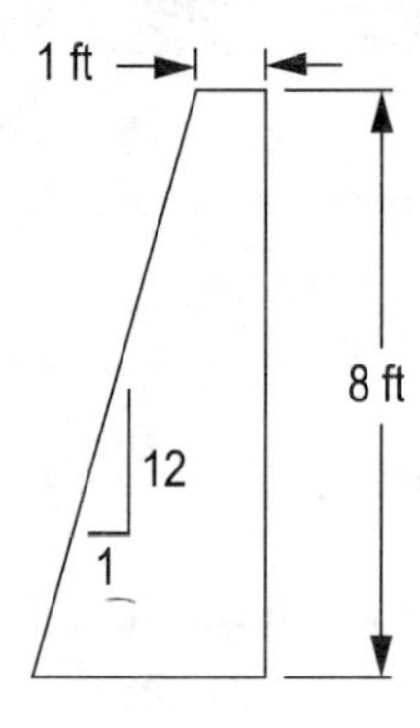

515. A bracket is anchored to a concrete wall using a bolt screwed into an imbedded insert as shown in the figure below. The tension (kips) in the bolt is most nearly:

(A) 0
(B) 4.5
(C) 12
(D) 32

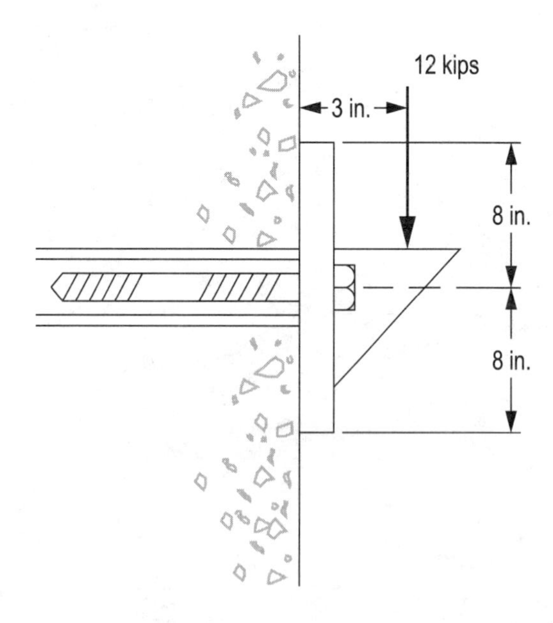

CONSTRUCTION AFTERNOON SAMPLE QUESTIONS

516. Federal OSHA standards require that supported scaffolds, including outrigger supports, be restrained from tipping when the height-to-base width ratio is more than:

(A) 2:1
(B) 3:1
(C) 4:1
(D) 5:1

517. Which statement is **NOT** true about the nuclear gages used for measuring asphalt and soil compaction?

(A) To be an authorized user of a nuclear gage an individual needs only to have a radiation-monitoring badge.

(B) Nuclear gages can be used to read moisture as well as density.

(C) It is required that a nuclear gage be transported in a properly labeled carrying case.

(D) Nuclear gages use low-level radioactive material.

518. A construction company has 750,000 employee hours worked with the following safety record:

Incidence Category	No. of Incidences
Minor injuries (first aid only)	10
Medical-only injuries (no lost time or light duty)	4
Medical injuries resulting in "light duty" restrictions	3
Lost-time injuries	5

The OSHA Incidence Rate is most nearly:

(A) 1.33
(B) 2.13
(C) 3.20
(D) 5.86

519. Laboratory testing was performed on a soil sample with the following results:

Sieve Analysis Data and Index Properties	
Sieve #	**% Passing**
3 in.	100
1 1/2 in.	98
3/4 in.	96
#4	77
#10	—
#20	55
#40	—
#100	30
#200	18
Liquid limit	32
Plastic limit	25

According to the Unified Soil Classification System, the classification of the sample is most nearly:

(A) SW
(B) SP
(C) SM
(D) SC

520. An end-bearing-on-rock pile foundation will be constructed to support a new bridge. The bottom of the pile cap will be at elevation 980. Pile splices will not be allowed. The pile embedment in the cap is 1′-0″. A maximum of a 2-ft cut-off due to driving damage is anticipated. Based on the subsurface exploration log shown, the minimum pile order length (ft) is most nearly:

(A) 50
(B) 55
(C) 60
(D) 65

SUBSURFACE EXPLORATION LOG

REGION 3
COUNTY ORANGE
PROJECT INTERSTATE 0
DATE START 5/8/07
DATE FINISH 5/9/07
CASING O.D. 2-1/2" I.D. ____
SAMPLER O.D. 2" I.D. 1-1/2"
RIG_TYPE ACKER B-40
CORE BARREL DOUBLE TUBE

HAMMER FALL-CASING 18"
HAMMER FALL-SAMPLER 30"
WEIGHT OF HAMMER-CASING 300 LBS.
WEIGHT OF HAMMER-SAMPLER 140 LBS.

HOLE BAF-3
LINE BASELINE
STA. 93+27
OFFSET 50' RT.
SURF. ELEV. 990.0

TIME	4 pm	8 am	2 pm
DATE	5/8/07	5/9/07	5/16/07
DEPTH TO WATER	6'	6'	6'

DEPTH BELOW SURFACE	BLOWS ON CASING	SAMPLE NO.	Blows on sampler 0–.5	.5–1.0	1.0–1.5	1.5–2.0	DESCRIPTION OF SOIL AND ROCK	MOIST. CONT. %
0	0	J1	1	0	1		BLACK MUCK WET - PLASTIC	115
	2						2'	
	11	J2	3	5	7			20
	25							
	31						GR SAND W/ ROOTS AND FIBERS	
	40							
	41	J3	8	8	9		MOIST - NON PLASTIC	8
	56							
	71							
	83						10'	
10	70	J4	6	5	5			29
	91							
	93							
	82						GR-BR CLAYEY SILT	
	93							
	81	J5	2	3	6		MOIST PLASTIC	31
	80							
	87							
	85							
	90						20'	
20	82	J6	4	3	3			34
	86							
	87						GR SILTY CLAY	
	85							
	90						MOIST - PLASTIC	
	73	J7	2	2	3			39
	72							
	83							
	71							
30	61							
	81	J8	2	2	2			40
	83							
	72							
	76							
	83							

THE SUBSURFACE INFORMATION SHOWN HEREON WAS OBTAINED FOR STATE DESIGN AND ESTIMATE PURPOSES. IT IS MADE AVAILABLE TO AUTHORIZED USERS ONLY THAT THEY MAY HAVE ACCESS TO THE SAME INFORMATION AVAILABLE TO THE STATE. IT IS PRESENTED IN GOOD FAITH, BUT IT IS NOT INTENDED AS A SUBSTITUTE FOR INVESTIGATIONS, INTERPRETATION OR JUDGMENT OF SUCH AUTHORIZED USERS.

CONTRACTOR ________ SM ________

DRILL RIG OPERATOR KLINEDINST
SOIL & ROCK DESCRIP. CHASSIE
REGIONAL SOILS ENGR. CHENEY
SHEET 1 OF 2
STRUCTURE NAME/NO. APPLE FREEWAY #2

29 HOLE BAF-3

SUBSURFACE EXPLORATION LOG

REGION 3
COUNTY ORANGE
PROJECT INTERSTATE 0
DATE START 5/8/07
DATE FINISH 5/9/07
CASING O.D. 2-1/2" I.D. ____
SAMPLER O.D. 2" I.D. 1-1/2"
RIG_TYPE ACKER B-40
CORE BARREL DOUBLE TUBE

HAMMER FALL-CASING 18"
HAMMER FALL-SAMPLER 30"
WEIGHT OF HAMMER-CASING 300 LBS.
WEIGHT OF HAMMER-SAMPLER 140 LBS.

HOLE BAF-3
LINE BASELINE
STA. 93+27
OFFSET 50' RT.
SURF. ELEV. 990.0

TIME	4 pm	8 am	2 pm
DATE	5/8/07	5/9/07	5/16/07
DEPTH TO WATER	6'	6'	6'

DEPTH BELOW SURFACE	BLOWS ON CASING	SAMPLE NO.	Blows on sampler 0–.5	.5–1.0	1.0–1.5	1.5–2.0	DESCRIPTION OF SOIL AND ROCK	MOIST. CONT. %
35	71	J9	2	3	2			36
	79						GR- SILTY CLAY	
	86							
	83						MOIST - PLASTIC	
	85							
40	82	J10	3	4	3			35
	81							
	93							
	91							
	96						45'	
	121	J11	20	21	35			10
	450						GR- SILTY GRAVEL	
	391							
	220						MOIST - NON PLASTIC	
	230							
50	200	J12	15	36	40		52' TO 53' CORED BOULDER	5
	370						RECOVERY 3"	
	400						MANY FRAGMENTS	
	410							
	380							
		J13	40	60	80			7
60							TOP OF ROCK 60.5'	
		J14	100	REFUSAL @			60.5'	
							HARD UNWEATHERED BASALT	
							RUN 1 60.5' TO 65.5' - 60" - RECOVERY 50", 12 PIECES RQD 70%	
							RUN 2 65.5' TO 70.5' - 60" - RECCVERY 60", 6 PIECES RQD 95%	
							HARD UNWEATHERED BASALT	
70								
							END OF BORING 70.5'	

THE SUBSURFACE INFORMATION SHOWN HEREON WAS OBTAINED FOR STATE DESIGN AND ESTIMATE PURPOSES. IT IS MADE AVAILABLE TO AUTHORIZED USERS ONLY THAT THEY MAY HAVE ACCESS TO THE SAME INFORMATION AVAILABLE TO THE STATE. IT IS PRESENTED IN GOOD FAITH, BUT IT IS NOT INTENDED AS A SUBSTITUTE FOR INVESTIGATIONS, INTERPRETATION OR JUDGMENT OF SUCH AUTHORIZED USERS.

CONTRACTOR ________ SM ________

DRILL RIG OPERATOR KLINEDINST
SOIL & ROCK DESCRIP. CHASSIE
REGIONAL SOILS ENGR. CHENEY
SHEET 2 OF 2
STRUCTURE NAME/NO. APPLE FREEWAY #2

30 HOLE BAF-3

After Cheney, R.S., and R.G. Chassie, *Soils and Foundations Workshop Manual*, National Highway Institute Course No. 13212, U.S. Department of Transportation, Federal Highway Administration, p. 224, November 1982.

GEOTECHNICAL
AFTERNOON SAMPLE QUESTIONS

This book contains 20 geotechnical depth questions, half the number on the actual exam.

GEOTECHNICAL AFTERNOON SAMPLE QUESTIONS

501. A soil specimen is obtained from below the groundwater table. The soil has a void ratio of 0.72 and a specific gravity of 2.65. The buoyant (submerged) unit weight (pcf) is most nearly:

(A) 60
(B) 76
(C) 96
(D) 122

502. A sample of a soft clay is required for consolidation testing. This sample is best obtained using a:

(A) dilatometer
(B) pressuremeter
(C) SPT
(D) Shelby tube

GO ON TO THE NEXT PAGE

503. The following data apply to a cylindrical specimen that was tested in the laboratory:

Sample diameter	3 in.
Sample length	6 in.
Sample weight before oven drying	2.95 lb
Sample weight after oven drying	2.54 lb
Specific gravity	2.65

The degree of saturation of the sample is most nearly:

(A) 60%
(B) 70%
(C) 80%
(D) 90%

504. A saturated cohesionless soil was tested to failure in a triaxial apparatus and the following data recorded:

Vertical total stress	33.5 psi
Chamber pressure	16.4 psi
Pore water pressure	10.0 psi

Under drained conditions, the effective friction angle is most nearly:

(A) 25°
(B) 30°
(C) 35°
(D) 40°

GO ON TO THE NEXT PAGE

505. A concrete dam with a sheet-pile cutoff wall is to be used to retain a 12-foot depth of water. The dam is 120 ft long, and the flow net is shown in the figure below. The granular soil has a coefficient of permeability of 0.003 fps (isotropic soil). The seepage loss (cfs) under the dam is most nearly:

(A) 0.014
(B) 1.400
(C) 1.700
(D) 17.000

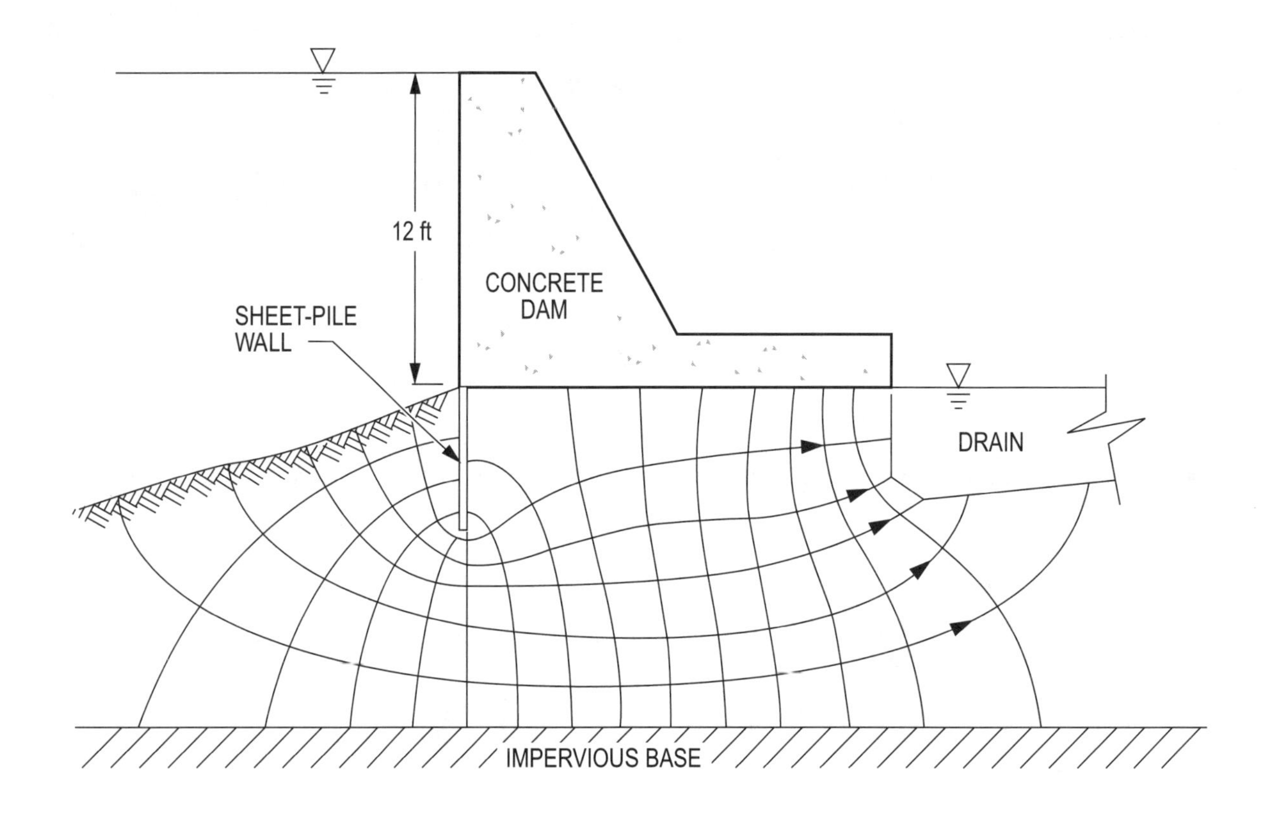

CONCRETE DAM SECTION

506. The following laboratory test data were obtained for a soil sample:

Sample volume	15.5 cm^3
Sample weight	30.5 g
Dry sample weight	25.0 g
Specific gravity	2.7

The void ratio and saturation of this sample are most nearly:

	Void Ratio	Saturation
(A)	0.4	0.9
(B)	0.7	1.0
(C)	0.4	1.0
(D)	0.7	0.9

507. A square mat foundation will be constructed at ground surface. The subsoil profile is shown in the figure below. The mat will have a uniform load of 500 psf. According to the chart on the following page, the primary consolidation settlement (in.) of the clay layer directly below the center of the mat is most nearly:

(A) 0.2
(B) 1.0
(C) 2.4
(D) 3.6

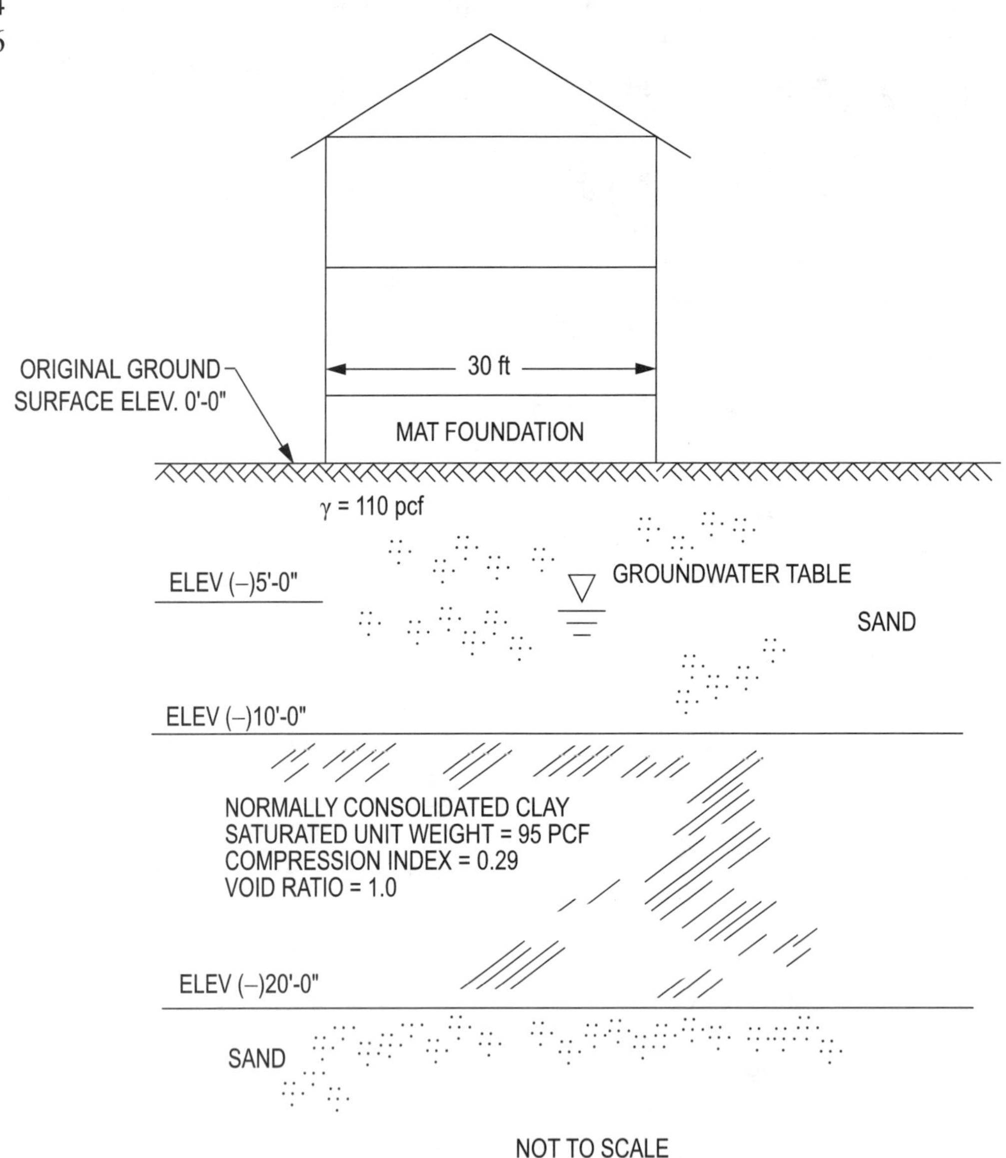

GO ON TO THE NEXT PAGE

507. (Continued)

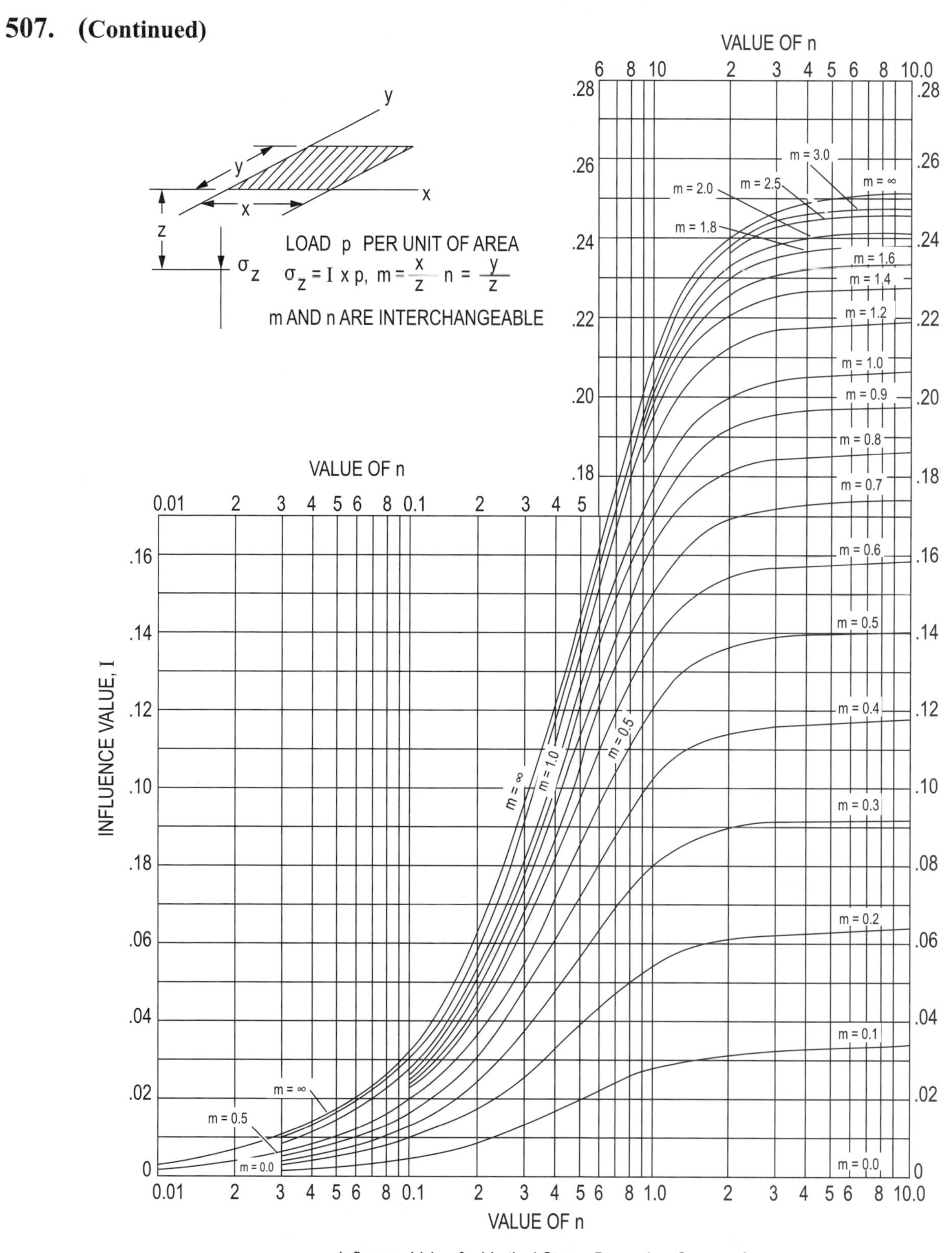

Influence Value for Vertical Stress Beneath a Corner of a Uniformly Loaded Rectangular Area (Boussinesq Case)

From *Soil Mechanics Design Manual 7.1*, Department of the Navy, Naval Facilities Engineering Command, Virginia, 1982.

GEOTECHNICAL AFTERNOON SAMPLE QUESTIONS

508. Structural fill is required below a slab-on-grade in an area with a deep frost line. To avoid potential problems with frost heave, the best material for structural fill would be:

(A) low-plasticity cohesive soil compacted dry of optimum (CL)

(B) inelastic silt (ML)

(C) silty sand (SM)

(D) well-graded sand (SW)

509. A 13-ft-long round timber pile with a 12-in.-diameter tip is to be driven into soil as shown in the figure below. From the information given, the ultimate end-bearing capacity (tons), using the Terzaghi method, is most nearly:

(A) 76
(B) 82
(C) 146
(D) 164

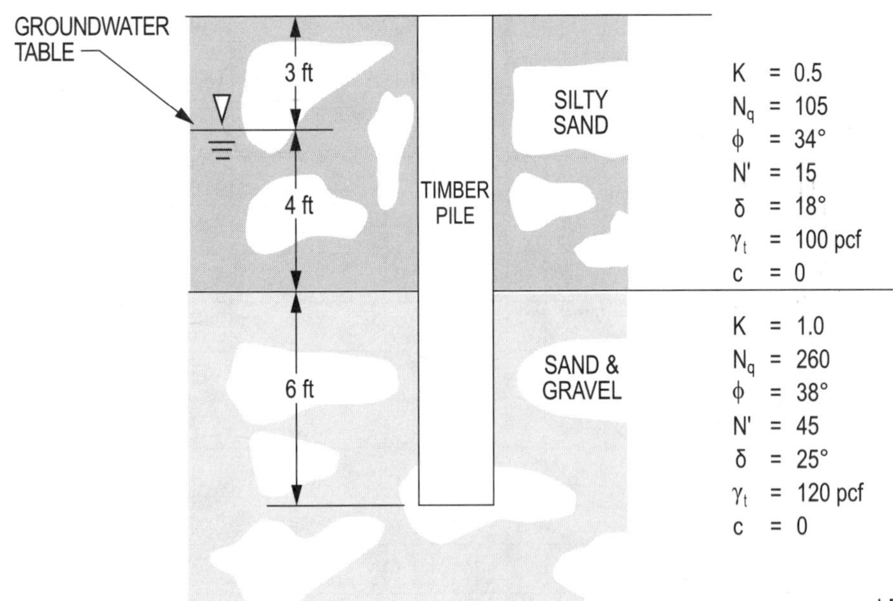

LEGEND

K = LATERAL EARTH PRESSURE COEFFICIENT
N_q = BEARING CAPACITY FACTOR
ϕ = SOIL FRICTION ANGLE
N' = STANDARD PENETRATION NUMBER (BLOWS PER FOOT) CORRECTED FOR EFFECTIVE OVERBURDEN PRESSURE
δ = SOIL-PILE INTERFACE FRICTION ANGLE
γ_t = SOIL TOTAL UNIT WEIGHT
c = COHESION

GO ON TO THE NEXT PAGE

510. A retaining system is proposed to protect a structure adjacent to an excavation. The retaining system should be sufficiently rigid so that only negligible deformations will occur during excavation. The most appropriate lateral earth pressure coefficient to use for design would be:

(A) K_a
(B) K_o
(C) K_p
(D) k_s

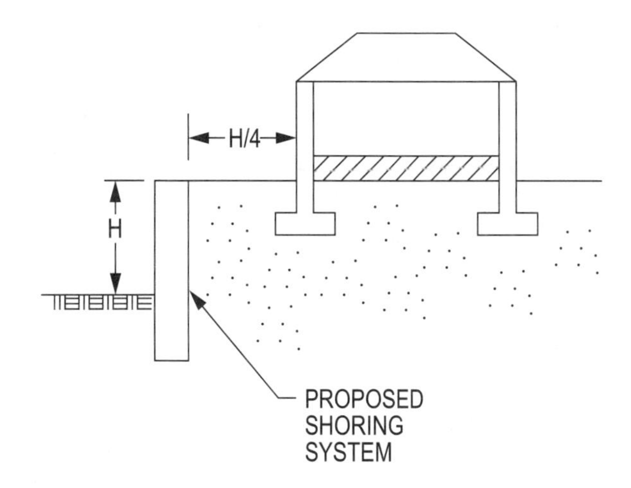

511. Refer to the soil profile shown. The effective vertical stress (psf) at Point A is most nearly:

(A) 1,200
(B) 1,400
(C) 1,500
(D) 1,700

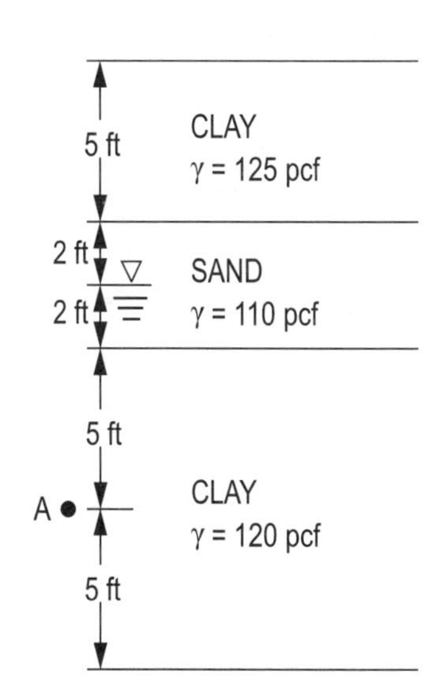

512. A 12-ft-high road embankment will be built using the reinforced earth method to resist lateral earth pressure as shown in the figure below. Assume the lateral earth coefficient of the sand backfill is 0.3, the friction angle between the reinforcing strip and the sand is 22°, and the required safety factor against slipping of the reinforcing strip is 3. The distance (ft) that the bottom reinforcing strip needs to be extended beyond the active failure surface is most nearly:

(A) 6
(B) 8
(C) 10
(D) 12

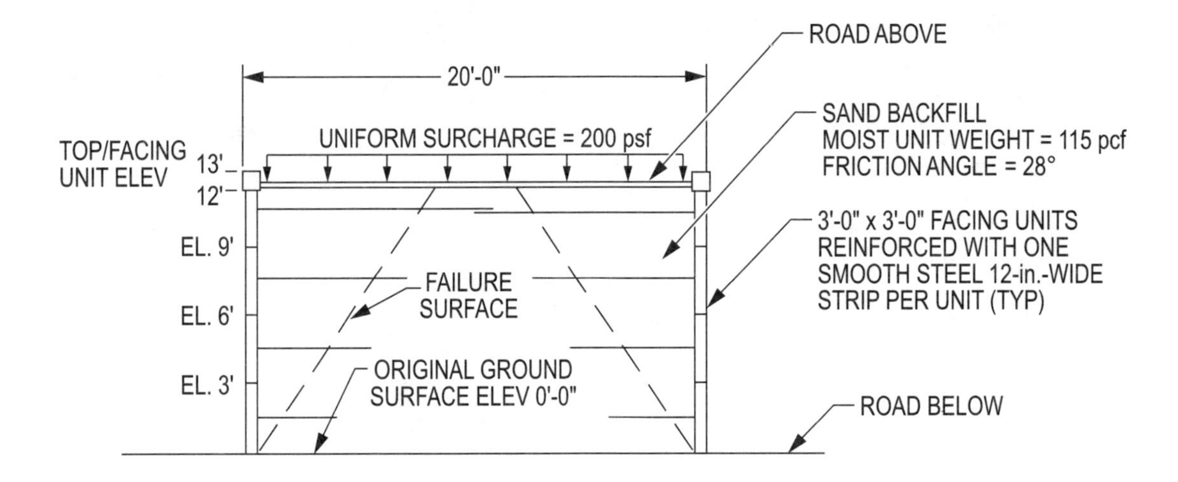

513. The liquefaction potential of a site is to be evaluated. The earthquake-induced average shear stress is 450 psf, and the cyclic stress ratio is 0.29. The factor of safety against liquefaction in Layer 3 is most nearly:

(A) 0.7
(B) 1.2
(C) 1.3
(D) 1.4

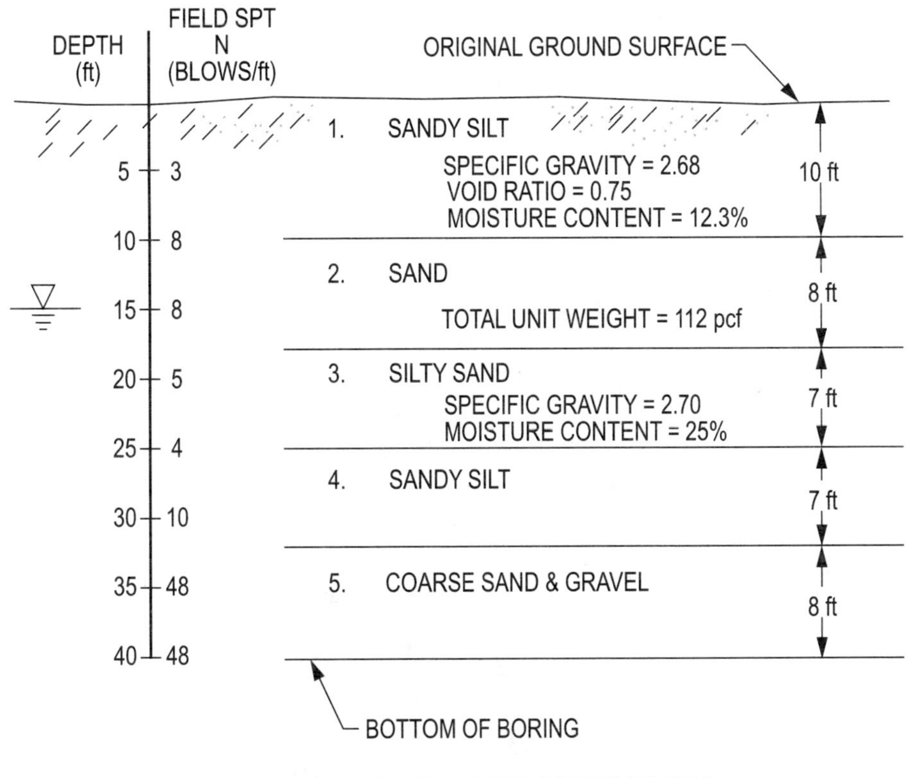

SUBSURFACE INVESTIGATION SUMMARY
NOT TO SCALE

GO ON TO THE NEXT PAGE

514. The geological conditions of a site are shown in the figure below. Well A is screened from 890–895 MSL. Well B is screened from 820–825 MSL. The static water level averages 895 MSL in Well A and averages 898 MSL in Well B. The formation between 820–840 MSL would be most accurately characterized as:

(A) a flowing artesian aquifer

(B) a deep water-table aquifer

(C) an interbedded lacustrine aquiclude

(D) a confined aquifer

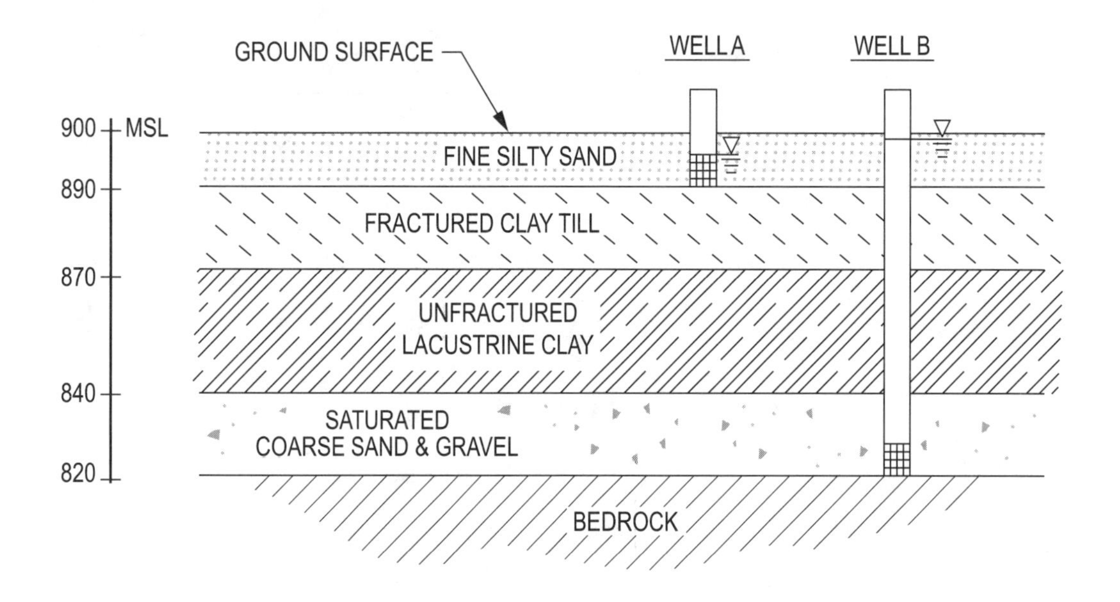

NOT TO SCALE

GO ON TO THE NEXT PAGE

515. A temporary slope will be excavated to the dimensions shown below. Laboratory testing has yielded the geotechnical parameters shown in the chart on the following page. If $\beta = 30°$, the safety factor for the failure surface shown is most nearly:

(A) 1.4
(B) 1.6
(C) 1.8
(D) 2.0

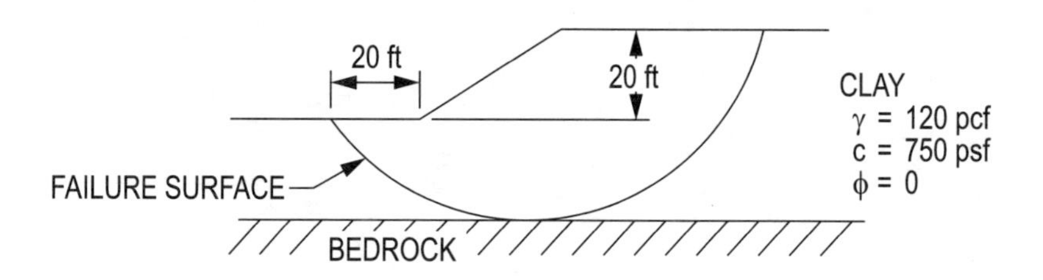

515. (Continued)

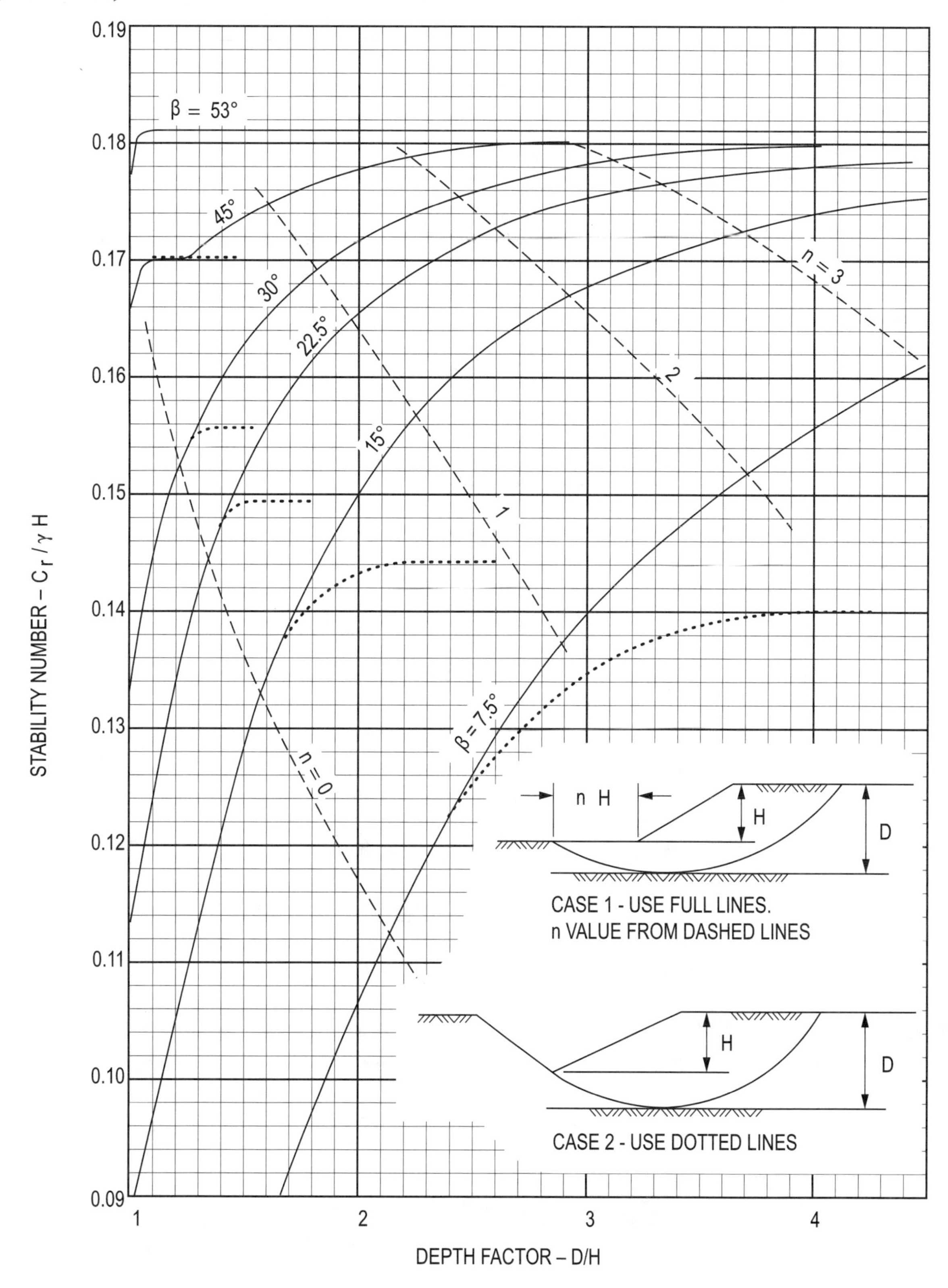

From *Fundamentals of Geotechnical Analysis,* Dunn, Anderson, Riefer, John Wiley and Sons, New York, 1980. Used by permission.

GO ON TO THE NEXT PAGE

516. The figure below shows the foundation and geotechnical data for a square footing. To achieve a safety factor of 3, the allowable bearing capacity (psf) is most nearly:

(A) 2,500
(B) 3,000
(C) 3,250
(D) 3,500

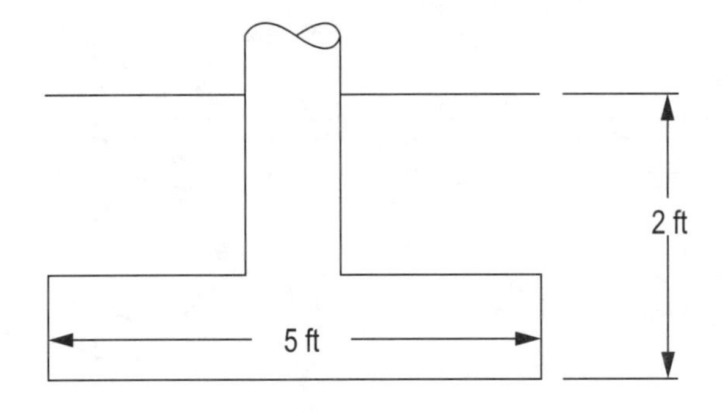

γ = 115 pcf
c = 200 psf
ϕ = 20°

ϕ	N_c	N_q	N_γ
0	5.14	1.0	0.0
5	6.5	1.6	0.5
10	8.3	2.5	1.2
15	11.0	3.9	2.6
20	14.8	6.4	5.4
25	20.7	10.7	10.8
30	30.1	18.4	22.4
32	35.5	23.2	30.2
34	42.2	29.4	41.1
36	50.6	37.7	56.3
38	61.4	48.9	78.0
40	75.3	64.2	109.4
42	93.7	85.4	155.6
44	118.4	115.3	224.6
46	152.1	158.5	330.4
48	199.3	222.3	496.0
50	266.9	319.1	762.9

517. A concrete gravity retaining wall is shown below. The safety factor against overturning is most nearly:

(A) 1.2
(B) 1.4
(C) 1.7
(D) 1.9

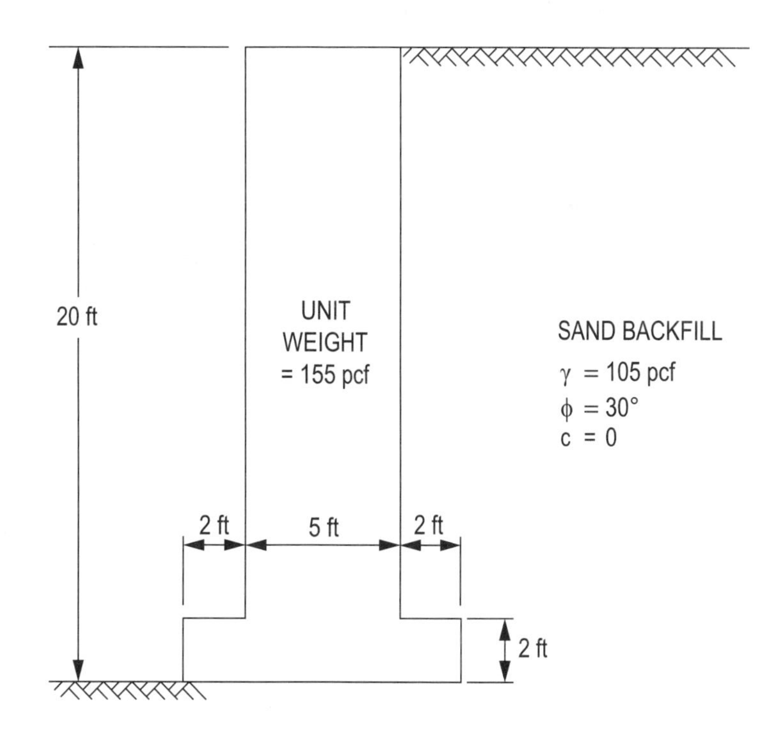

518. A 2-ft × 2-ft square concrete precast pile is shown below. The concrete unit weight is 150 pcf. Ignore the resistance of the soft clay but include the weight of the pile. The interface soil friction angle on the pile is 0.75 ϕ. For a safety factor of 3.0, the allowable tensile capacity (kips) of the pile is most nearly:

(A) 50
(B) 100
(C) 150
(D) 200

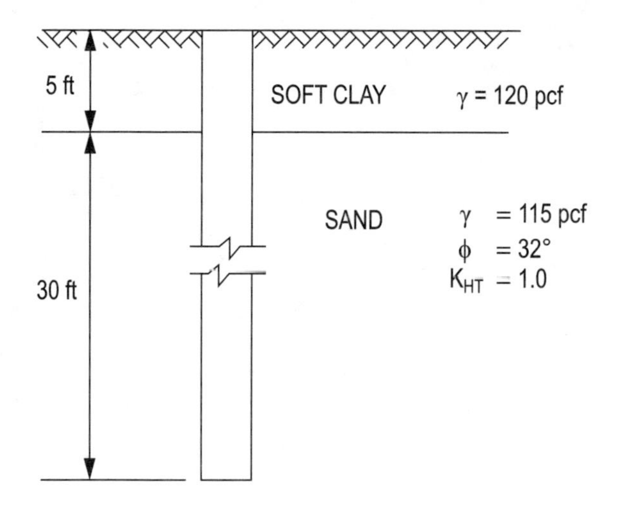

519. An anchored sheet pile sea wall is shown below. Which of the following will **NOT** improve the factor of safety against failure?

(A) An increase in the relative density of the silty sand layer in front of the wall

(B) Extending the wall embedment an additional 2 ft

(C) A sudden water elevation drop in the channel

(D) A gradual water surface elevation drop in the channel

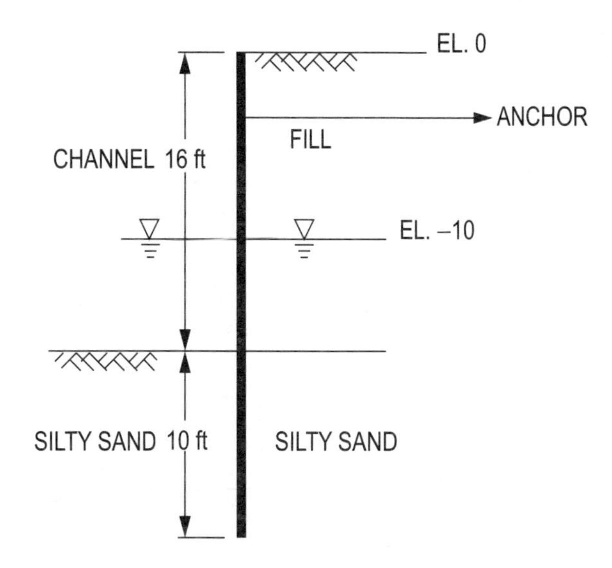

520. A segment of interstate highway requires the construction of an embankment of 500,000 yd^3. The embankment fill is to be compacted to a minimum of 90% of Modified Proctor maximum dry density.

A source of suitable borrow has been located for construction of the embankment. Assume that there is no soil loss in transporting the soil from the borrow pit to the embankment. The following data apply:

Specific gravity of the soil particles	2.65
Modified Proctor maximum dry density	120 pcf

Assuming each truck holds 5.0 yd^3 and the void ratio of the soil is 1.30 during transport, the minimum number of truckloads of soil from the borrow pit that is required to construct the embankment is most nearly:

(A) 100,000
(B) 150,000
(C) 200,000
(D) 250,000

STRUCTURAL
AFTERNOON SAMPLE QUESTIONS

This book contains 20 structural depth questions, half the number on the actual exam.

501. You may select **EITHER** the ASD **OR** the LRFD option.

A 7-ft-long W10 × 39 steel beam supports a moving concentrated load as shown in the figure below. The load can be applied at any point along the beam. Assuming the beam is laterally supported, the maximum load (kips) that can be applied to the beam is most nearly:

	ASD	LRFD
(A)	63	94
(B)	67	101
(C)	71	107
(D)	125	187

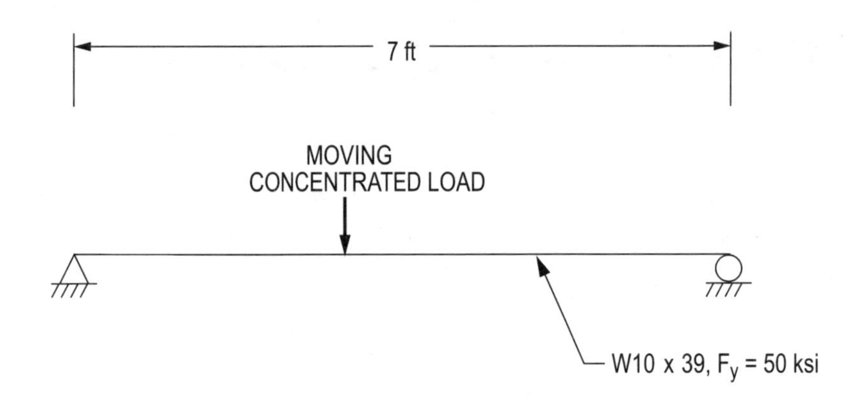

502. The following data apply to the structure shown below.

Ground snow load = 20 psf.
Roof is fully exposed with wood shingles.
Open terrain
Occupancy Category I
Unheated structure

Design Code:
ASCE 7, *Minimum Design Loads for Buildings and Other Structures,* 2005.

Roof joists span from the exterior walls to the ridge beam. The design snow load (psf) for roof joists is most nearly:

(A) 10
(B) 12
(C) 16
(D) 20

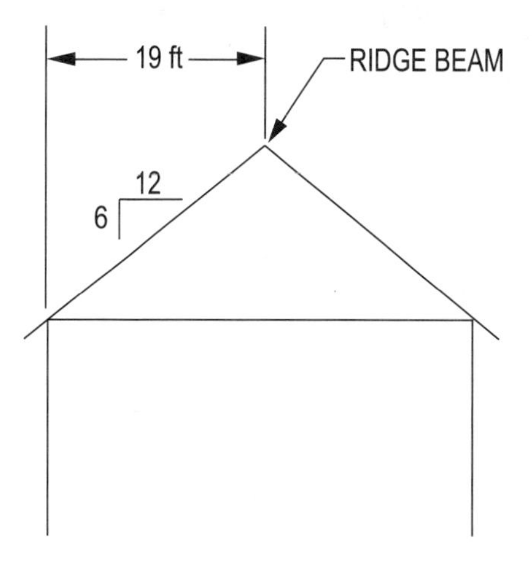

503. An elevated water tank is supported by a four-legged braced-steel tower frame, as shown in the two following figures.

Assumptions:

Tank is either completely empty or completely full of water, for load combinations.

Full-tank water volume is 30 ft in diameter × 20 ft deep.

Top of pedestal is at the grade elevation.

Ignore wind perpendicular to the top of the tank.

No snow load

Design Data:

Weight of empty tank	70 kips
Weight of tower frame	400 lb/linear foot of height
Weight of backfill	100 pcf
Normal weight concrete	150 pcf

If the total wind force perpendicular to the side of the tank is 30 kips and the wind force on the tower frame is neglected, the maximum net uplift (kips) at the top of a pedestal is most nearly:

(A) 250
(B) 75
(C) 50
(D) 25

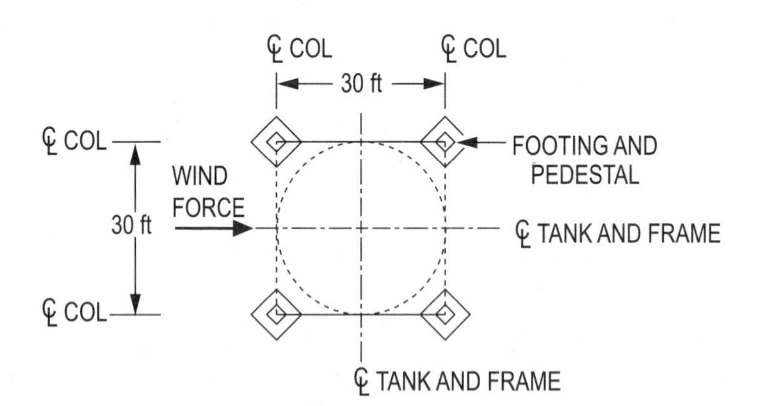

TOWER COLUMN LOCATION PLAN
NOT TO SCALE

503. (Continued)

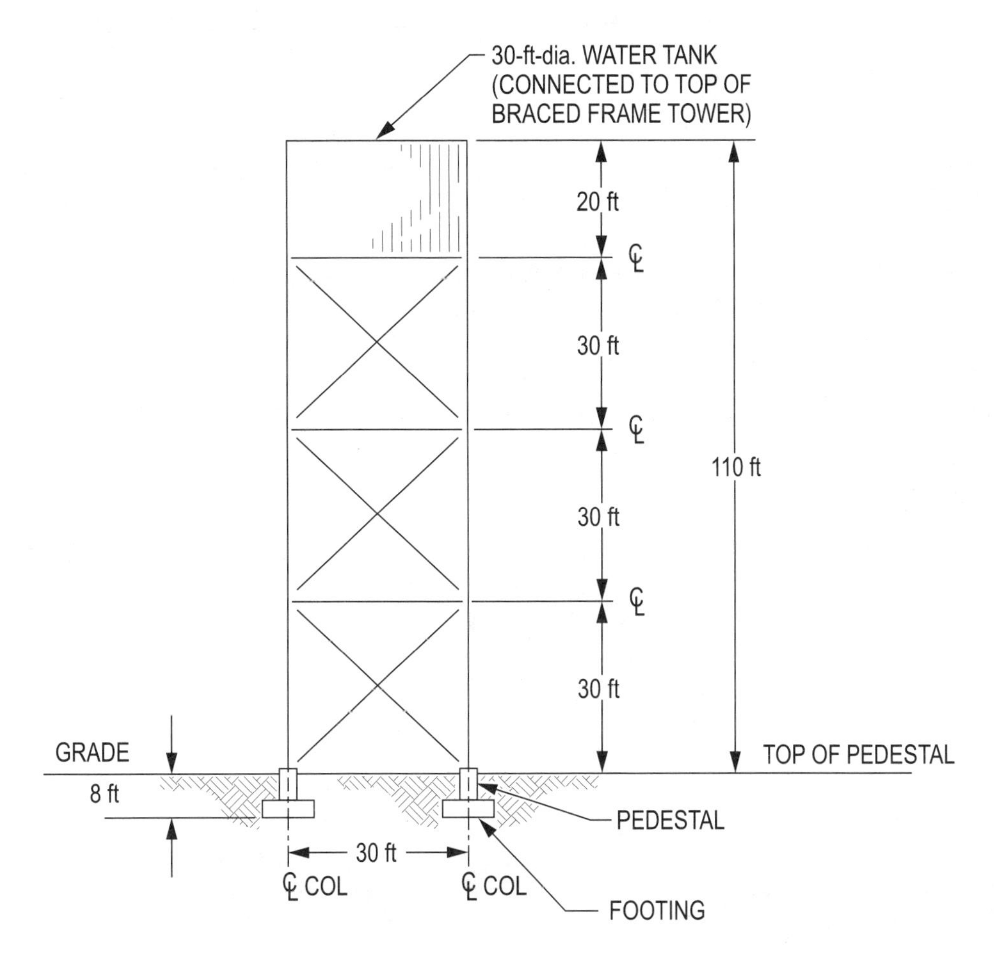

TYPICAL TOWER FRAME ELEVATION
NOT TO SCALE

GO ON TO THE NEXT PAGE

504. The required factored moment for the beam shown below is 80 ft-kips. The following data apply:

f'_c = 4 ksi
f_y = 60 ksi

Design Code:
ACI 318, *Building Code Requirements for Structural Concrete,* 2005.

The required minimum reinforcement A_s (in^2) for the factored moment is most nearly:

(A) 1.10
(B) 1.04
(C) 0.88
(D) 0.66

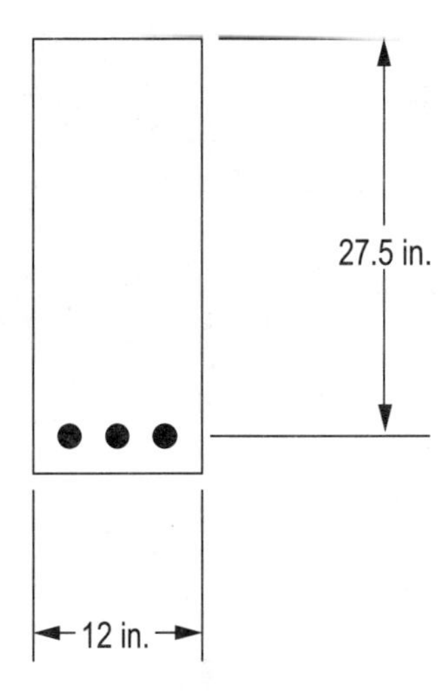

505. The following data apply to the masonry beam section below:

$f'_m = 1,500$ psi

$f_y = 60,000$ psi

Design Code:
ACI-530/530.1, Building Code Requirements and Specifications for Masonry Structures and Related Commentaries, 2005.

Use allowable stress design.

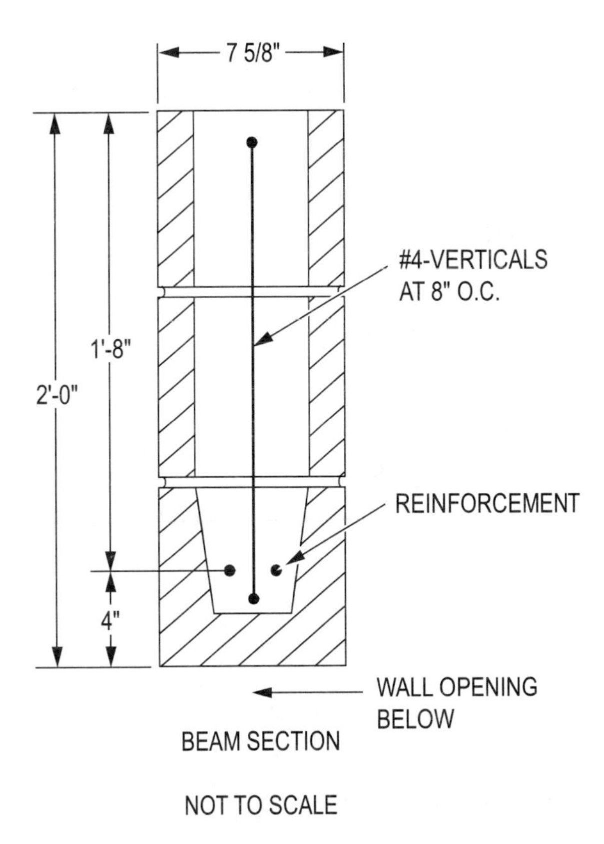

BEAM SECTION

NOT TO SCALE

The maximum allowable shear force (kips) for the beam is most nearly:

(A) 6
(B) 12
(C) 14
(D) 18

GO ON TO THE NEXT PAGE

506. The figure on the following page shows connections for the diagonal to chord members of a truss. Truss members are steel double angles. The truss carries gravity loads only.

Assumptions:

Members are axially loaded through their centroids.
Members have intermittent fillers and connection spacing, which satisfies the AISC requirements for built-up members.
Bolts are on a standard gage.

Design Code:

AISC, *Steel Construction Manual*, 13th edition, 2005 (without supplements).

Design Data:

Structural steel is ASTM A36 $(F_y = 36 \text{ ksi and } F_u = 58 \text{ ksi})$.
Bolts are ASTM A325N.
Bolt holes are standard.

Refer to the top chord to diagonal connection in the figure on the following page.

Use **EITHER** the ASD **OR** the LRFD provisions.

For the tension force in the figure, the size of the least weight double-angle diagonal is most nearly:

(A) ⅃L 2 1/2 × 2 1/2 × 3/8

(B) ⅃L 3 × 3 × 3/8

(C) ⅃L 3 1/2 × 3 1/2 × 5/16

(D) ⅃L 4 × 4 × 3/8

506. (Continued)

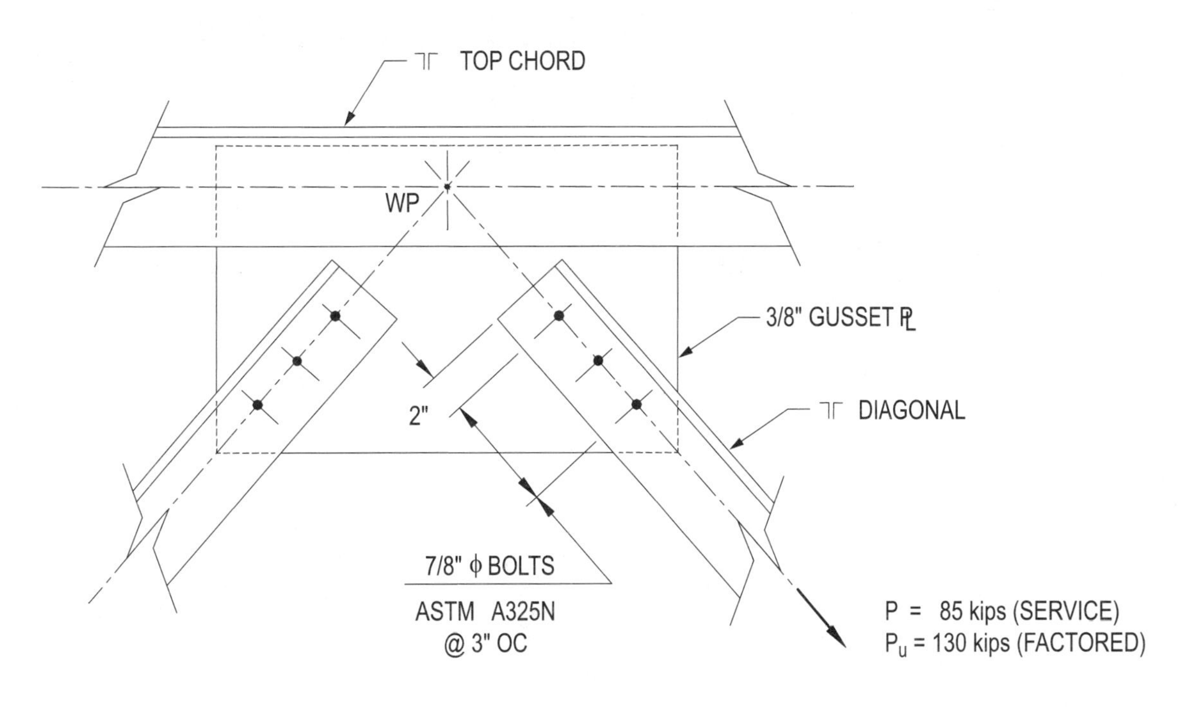

NOT TO SCALE

507. The following assumptions apply to the building shown below:

1. There is only shear transfer between the two adjacent diaphragms.
2. Neglect bending deflection of the diaphragms.
3. Shear deflection of a cantilevered diaphragm $\Delta_V = \dfrac{VL}{bG'}$

The shear load (kips) transferred to Shear Wall A is most nearly:

(A) 26
(B) 36
(C) 40
(D) 64

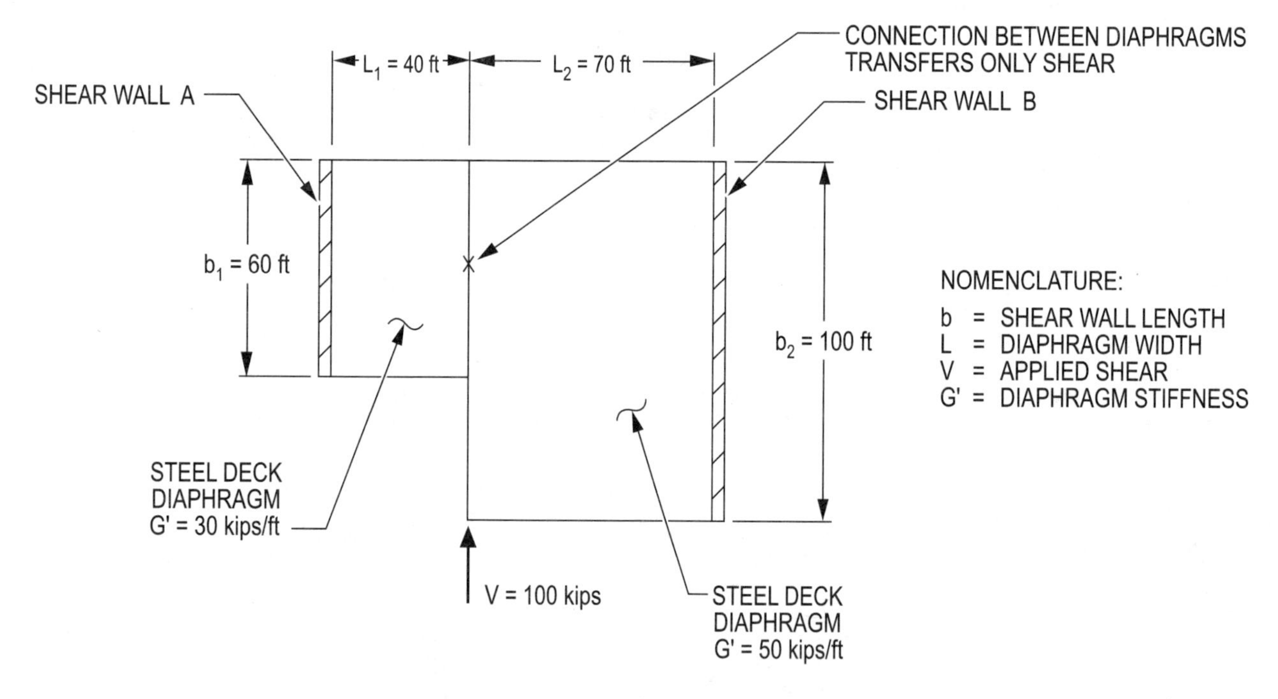

508. You may select **EITHER** the ASD **OR** the LRFD option.

As shown in the figure, a 3/4-in. × 9-in. plate is subject to a shear force of 35 kips (ASD) or 45.5 kips factored (LRFD) and a tensile force due to wind load. The plate is welded to a 3/4-in. plate with 1/4 E70XX weld. Ignore any eccentricity. Do not use the alternate provisions of J2.4 a, b, or c.

The tension force capacity (kips) for these welds is most nearly:

ASD (service, without 1/3 stress increase for wind)

(A) 37.4
(B) 63.4
(C) 72.4
(D) 212.6

LRFD (ultimate)

(A) 63.1
(B) 98.6
(C) 108.6
(D) 212.6

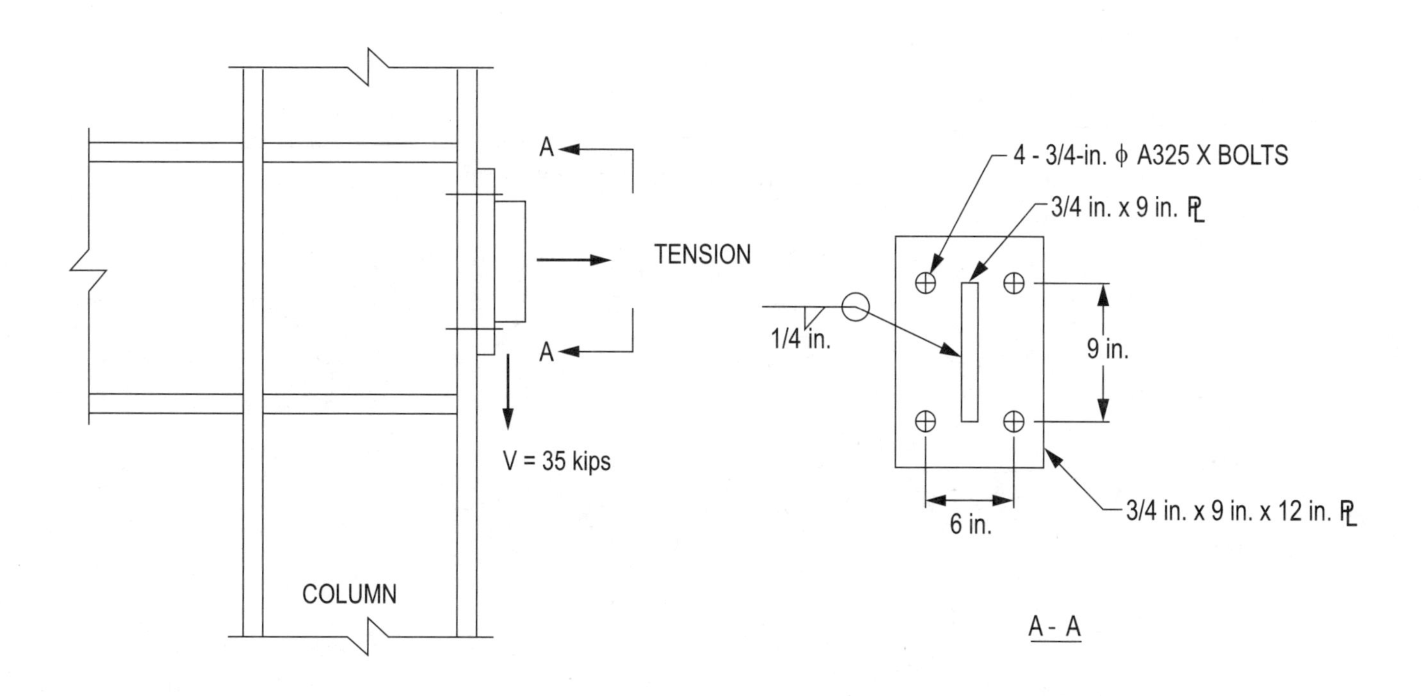

509. The figure on the following page shows a simply supported prestressed concrete double-tee spanning 60 ft between supports.

Design Code:
ACI 318, *Building Code Requirements for Structural Concrete,* 2005.

Design Data:
Double-Tee Properties:
Cross-sectional area, $A_c = 410 \text{ in}^2$
Concrete unit weight = 110 pcf

Prestress Strand Properties:
(8) 1/2-in.-ϕ low-relaxation strands (4 per stem)
Area, $A_{ps} = 0.153 \text{ in}^2$ per strand
Ultimate stress, $f_{pu} = 270{,}000$ psi
Initial stress immediately after release, $f_{pi} = 195{,}000$ psi
Eccentricity at midspan, $e_m = 14.25$ in. (harped at midspan)
Eccentricity at ends of span, $e_e = 9.50$ in.

Assumptions:
Positive (+) moments produce tension stresses in the bottom concrete fiber.
Prestressing strands are fully bonded throughout the member.

The bending moment diagram, due to the combined effect of initial prestress and the double-tee weight, is most nearly:

(A) 

(B) 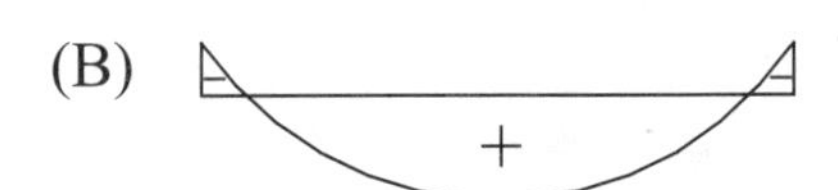

(C) 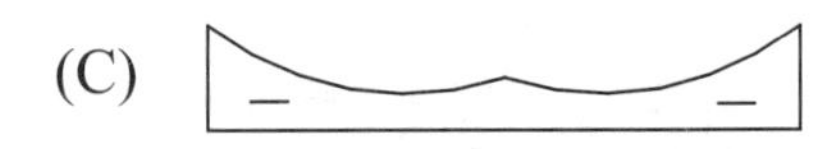

(D) 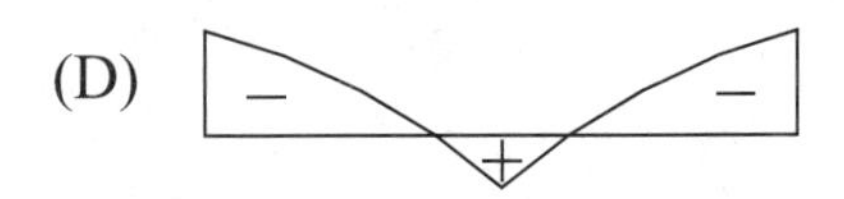

GO ON TO THE NEXT PAGE

509. (Continued)

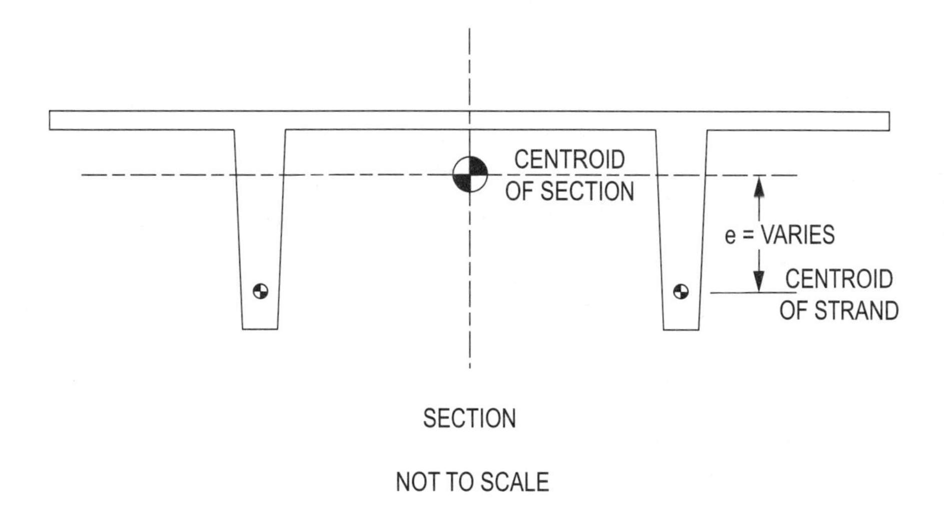

510. Two columns are supported by a reinforced concrete combined footing as shown in the figure.

Given:
- Neglect self-weight of the footing.
- Neglect earth cover on footing.
- Columns are 24 in. square.
- The total footing thickness is 3'-4".

Average effective depth d = 36 in. to the bottom reinforcing bars in the footing.
$P_1 = P_2 = 300$ kips (no eccentricity)

Design Code:
ACI 318, *Building Code Requirements for Structural Concrete*, 2005.

With respect to bending about the footing's Y-axis, the critical service shear force (kips) is most nearly:

(A) 109
(B) 150
(C) 191
(D) 218

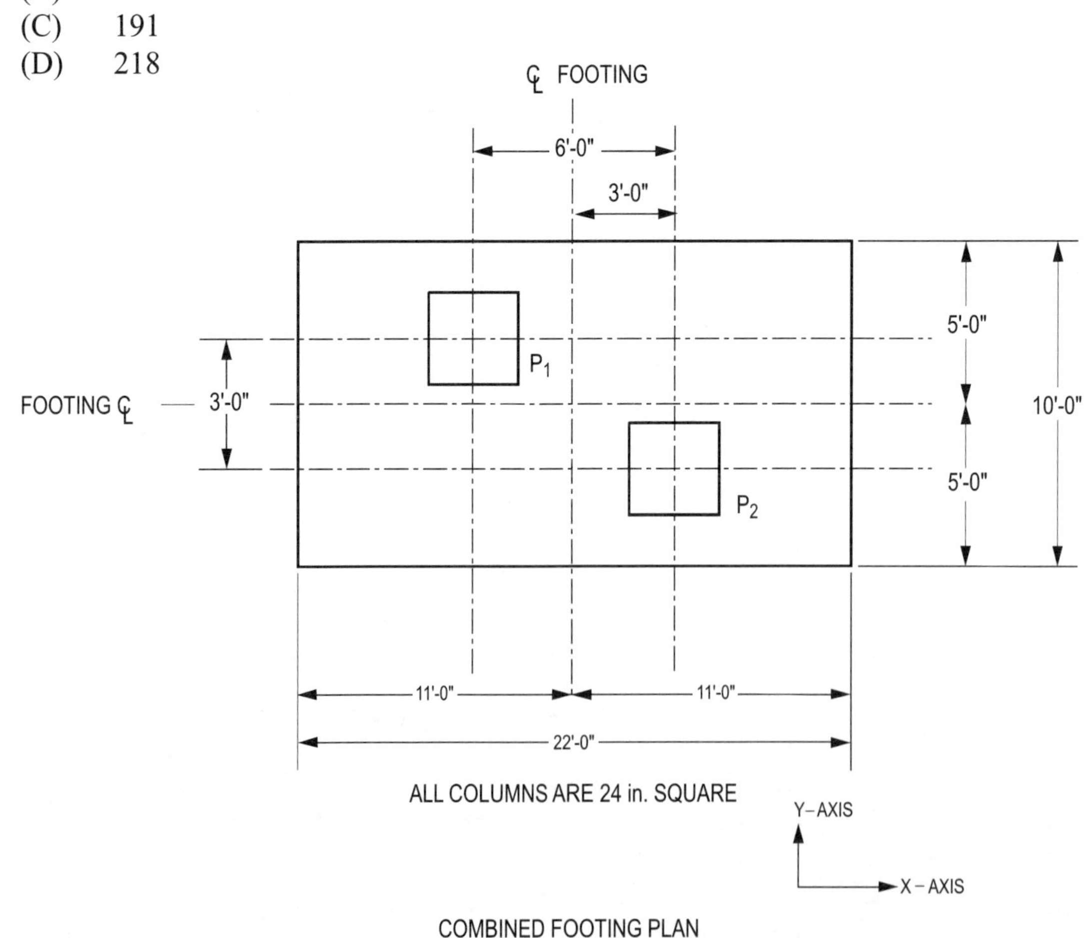

COMBINED FOOTING PLAN
NOT TO SCALE

GO ON TO THE NEXT PAGE

511. The following information applies to the masonry structure shown in the figure.

Design Code:
ACI 530, *Building Code Requirements for Masonry Structures,* 2005.

Materials:
Hollow concrete masonry units $f'_m = 1,500$ psi with Type S mortar. Cells with reinforcing are grouted.

Steel reinforcement ASTM A615 Grade 60

Loads:
Roof dead load = 30 psf
Average wall dead load = 54 psf
Design wind (pressure or suction) = 20 psf
Seismic forces do not govern.

Use allowable stress design.

The maximum design moment (ft-lb/ft) for the masonry wall for dead and wind loads is most nearly:

(A) 250
(B) 360
(C) 480
(D) 580

GO ON TO THE NEXT PAGE

511. (Continued)

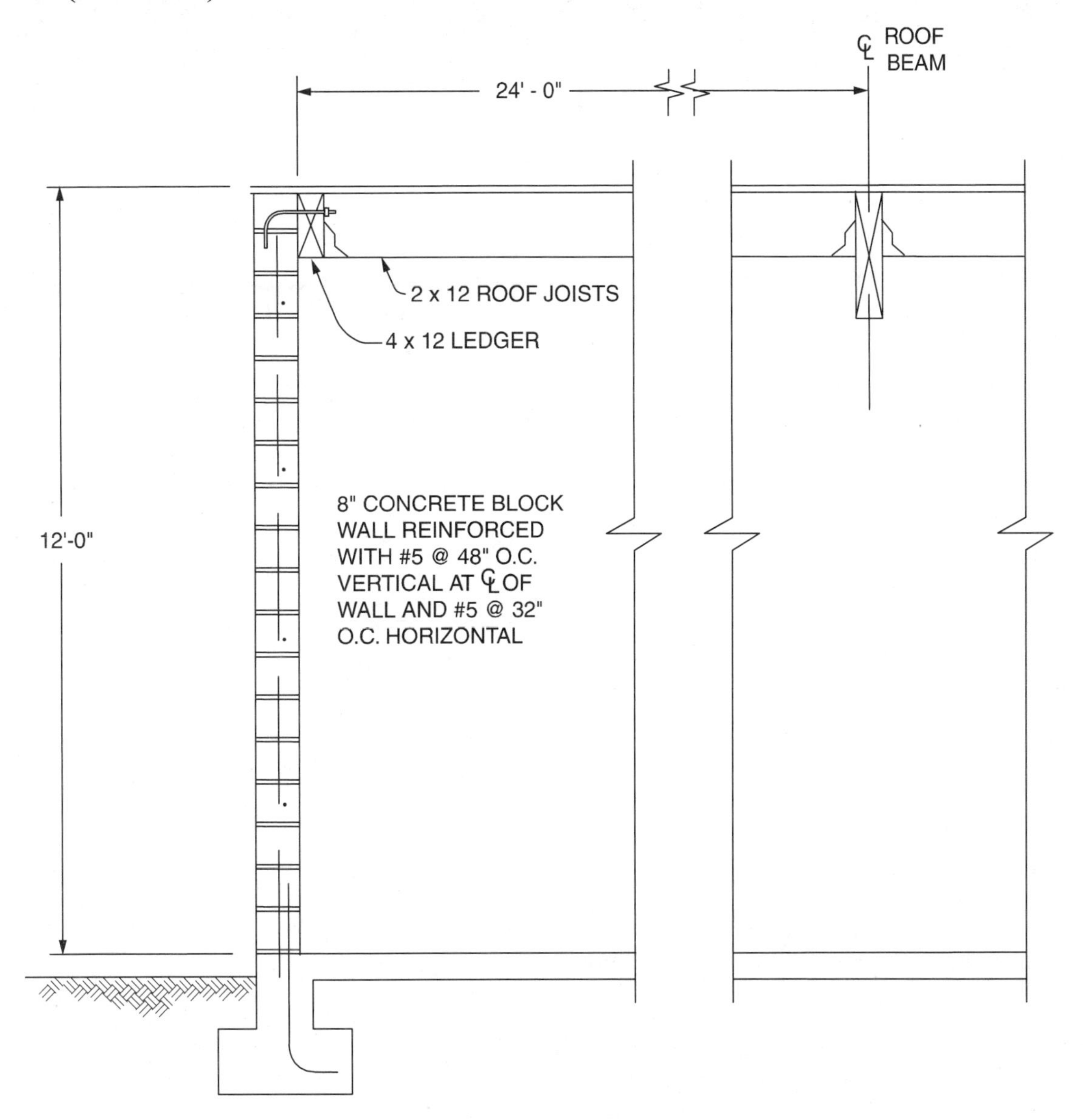

GO ON TO THE NEXT PAGE

512. Two solid aluminum shafts are rigidly attached together and attached to a rigid support at one end. Assume G = 4,400 ksi. Two torques are applied as shown. The angle of twist at End B (radians) is most nearly:

(A) 0.2
(B) 0.5
(C) 0.6
(D) 0.7

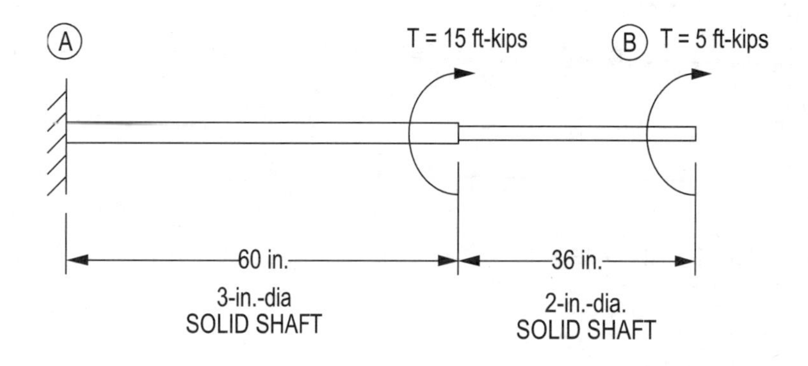

513. For the section shown below, the unit shear force (kips/in.) in the welds connecting the top flange to the web (total unit shear in the two weld lines combined) is most nearly:

(A) 1.5
(B) 3.0
(C) 5.5
(D) 10.0

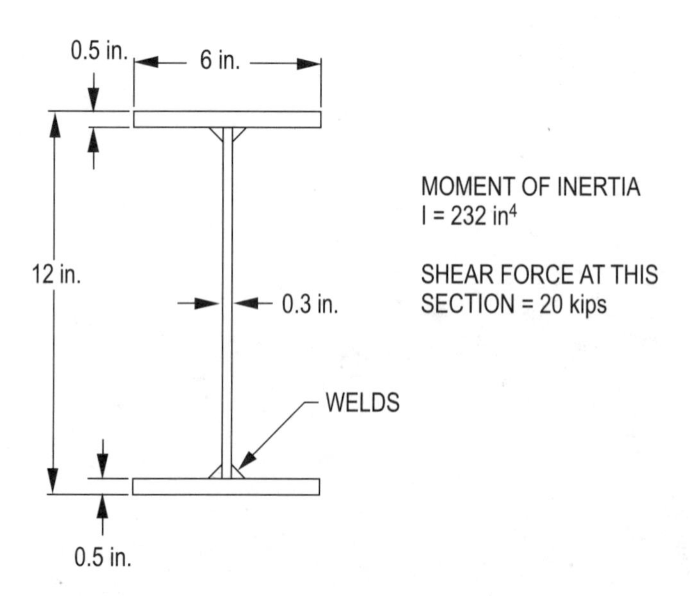

GO ON TO THE NEXT PAGE

514. Full sawn 2 × 12 beams span across supports and are loaded as shown below.

Design Code:
National Design Specification for Wood Construction, 2005 ASD edition & *National Design Specification Supplement*, 2005 ASD edition.

The maximum shear stress (psi) parallel to the grain is most nearly:

(A) 35.6
(B) 33.8
(C) 30.0
(D) 20.0

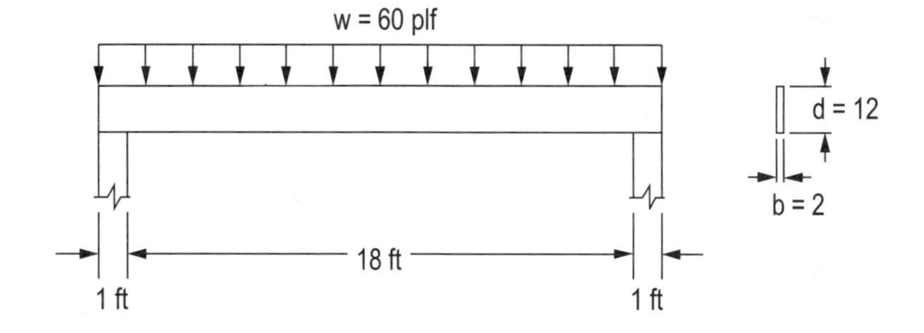

515. Design Code:
National Design Specification for Wood Construction, 2005 ASD edition & *National Design Specification Supplement*, 2005 ASD edition.

Assuming the member is not incised, the allowable design value F'_{Θ} (psi) for the roof member bearing is most nearly:

(A) 452
(B) 499
(C) 630
(D) 1,004

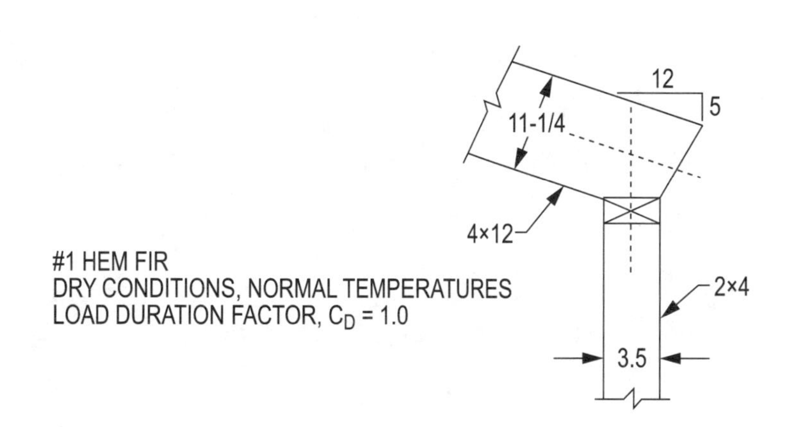

516. The cross section for a 50-ft-span, rectangular, prestressed beam is shown below. The beam has no mild reinforcing steel.

Design Code:
PCI Design Handbook, 6th edition, 2004.

Design data for prestressing strands:
Low relaxation – 1/2-in. diameter

$f_{pu} = 270$ ksi

$A_s = 0.153$ in^2 per strand

Stress at release = 175 ksi per strand (after initial losses)

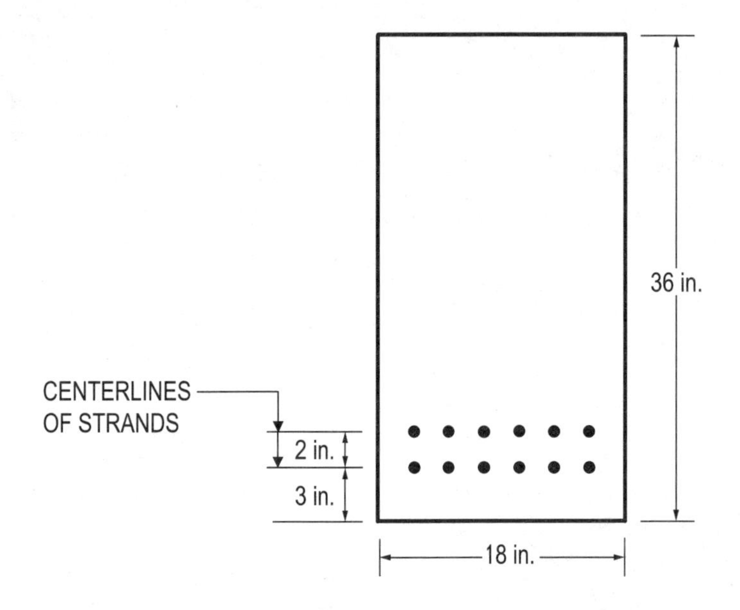

If the top fiber stress at the mid-span of the beam due to the beam self-weight is 0.65 ksi, the total top fiber stress (ksi) at release is most nearly:

(A) 0.01 (tension)

(B) 0.50 (tension)

(C) 0.99 (compression)

(D) 2.29 (compression)

517. The figure below shows a cross section of a deck slab of a steel girder bridge.

Design Code:
AASHTO LRFD Bridge Design Specifications, 3rd edition, 2004, with 2005 and 2006 Interim Revisions.

The effective span length S for the deck slab is most nearly:

(A) 6'-6"
(B) 7'-0"
(C) 7'-6"
(D) 8'-6"

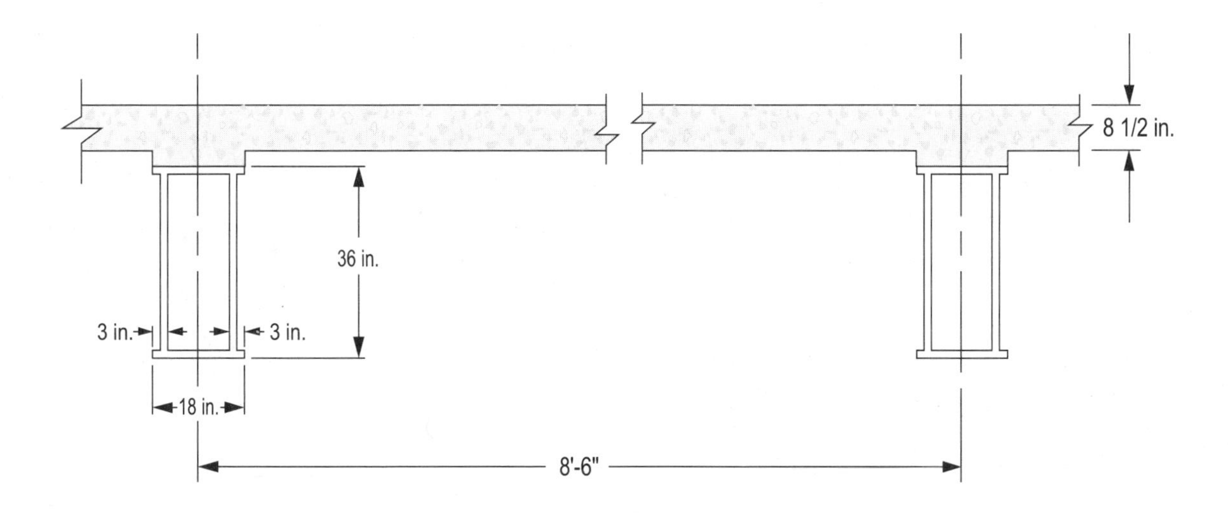

518. A crane runway steel-rolled beam is being used approximately 48 times each day. Each application involves reversal of tensile or compressive stress.

Design Code:
AISC *Steel Construction Manual*, 13th edition.

Assuming adequate basic allowable stress, the design stress range (ksi) for the beam during 25 years of service is most nearly:

(A) 16
(B) 24
(C) 29
(D) 37

519. The concrete footing shown below is subject to loads from a braced frame. Assume the footing is rigid compared to the soil. The maximum bearing pressure (ksf) under the footing is most nearly:

(A) 0.5
(B) 1.0
(C) 2.0
(D) 2.7

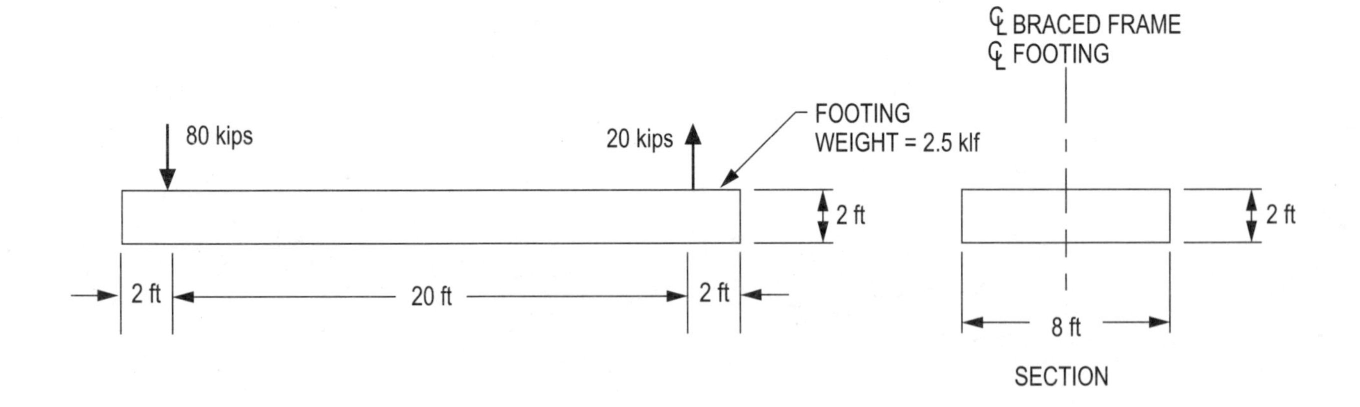

GO ON TO THE NEXT PAGE

520. Neglecting the tip capacity, the shortest pile length L (ft) required to achieve a design capacity of 50 kips and a factor of safety of 2.5 is most nearly:

(A) 15
(B) 18
(C) 30
(D) 38

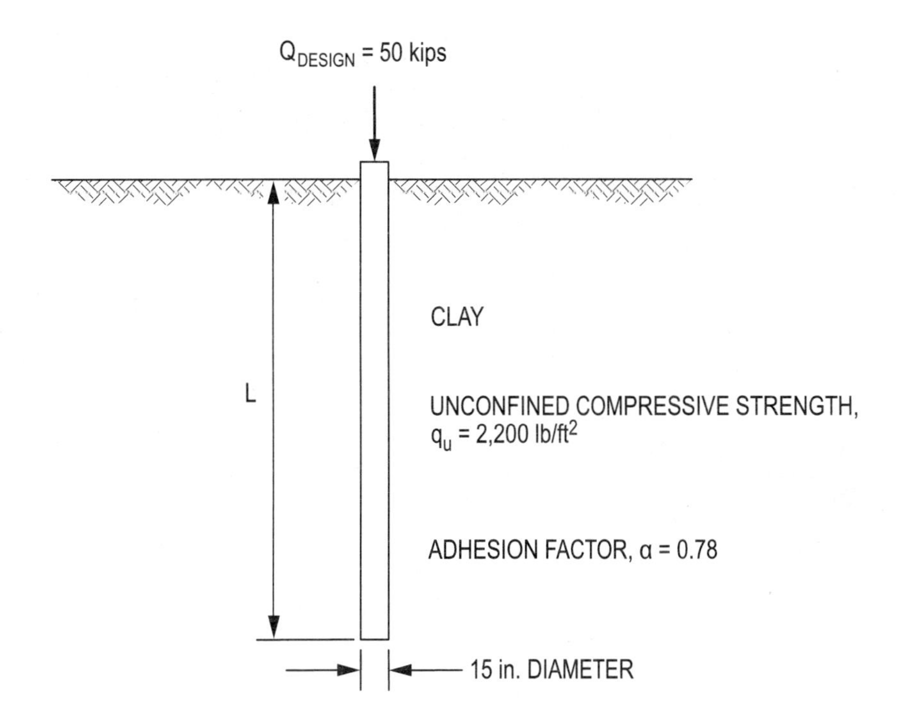

TRANSPORTATION
AFTERNOON SAMPLE QUESTIONS

This book contains 20 transportation depth questions, half the number on the actual exam.

501. The signalized intersection shown in the figure below has a pedestrian demand of 0.9 pedestrians per cycle and an average pedestrian speed of 4.0 fps. The minimum green time (sec) for pedestrians traveling in the North-South direction is most nearly:

(A) 22
(B) 27
(C) 29
(D) 32

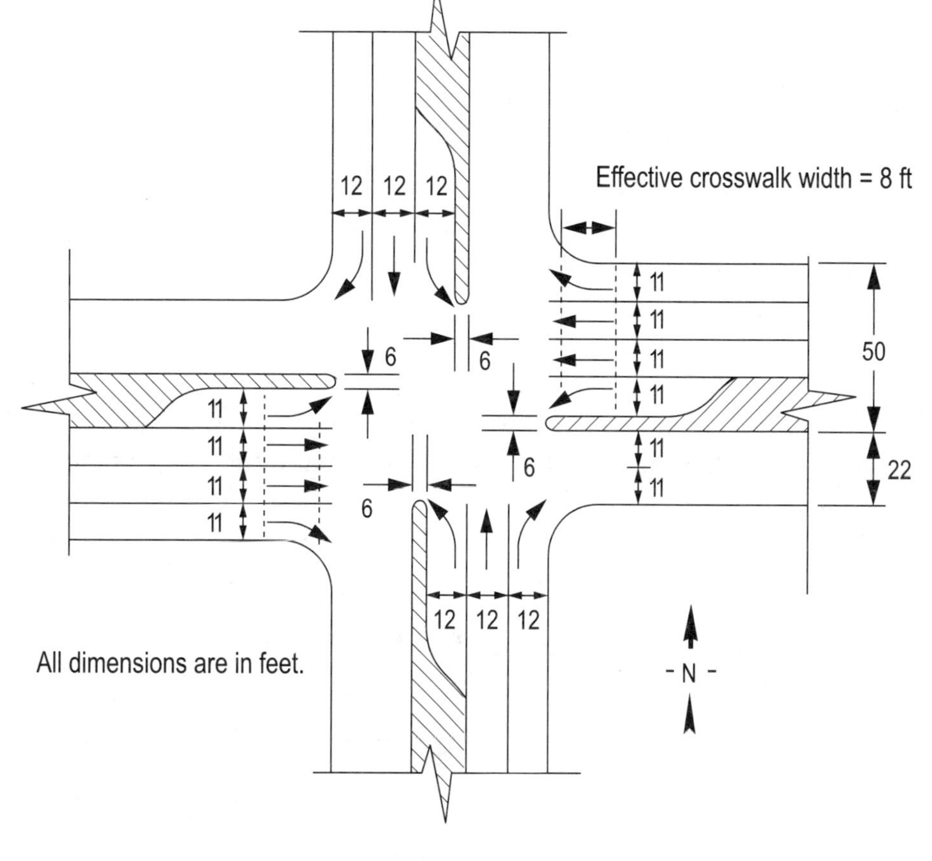

INTERSECTION PLAN
NOT TO SCALE

502. A freeway has a free-flow speed of 70 mph. The maximum service flow rate per lane at Level of Service D (passenger cars per hour per lane, or pcphpl) is most nearly:

(A) 1,855
(B) 1,980
(C) 2,150
(D) 2,400

503. A traffic engineering study is being conducted on an urban four-lane arterial street. An automatic traffic counter using a pneumatic tube detector gave a total 1-year count of 24,560,000 axles at a continuous count station. Manual classification studies at the count station indicate that the traffic stream consisted of 85% passenger cars, 10% three-axle trucks, 3% four-axle trucks, and 2% five-axle trucks.

The average annual daily traffic (AADT) volume (vehicles) at this count station is most nearly:

(A) 2,800
(B) 25,770
(C) 27,400
(D) 30,310

504. The vertical alignment of a highway is shown in the figure below. The stopping sight distance (ft) is most nearly:

(A) 440
(B) 490
(C) 525
(D) 650

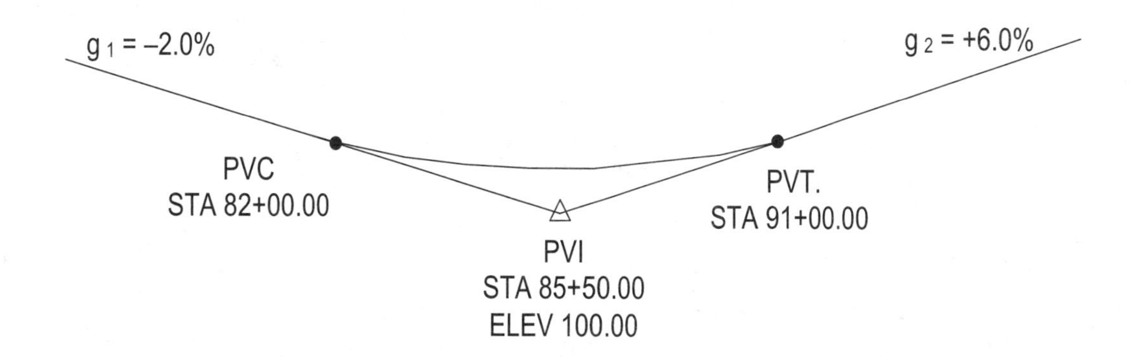

505. Consider the relocation and construction of a section of rural highway. Data on costs are given in the table below. All values are in thousands of dollars. Use a 20-year analysis period with a 10% annual interest rate. Major maintenance will **NOT** be done in the 20th year.

First cost	\$6,000
Annual maintenance for first 10 years	\$50
Annual maintenance for second 10 years	\$75
Major maintenance every 10 years	\$300
Residual value	\$3,000
Annual road user costs	\$660

The present value or present worth of the highway costs, ignoring user costs, is most nearly:

(A) \$6,000,000
(B) \$6,150,000
(C) \$6,350,000
(D) \$6,600,000

506. If the traffic volume on a highway in 2006 is 30,000 vehicles per day, with a predicted annual growth rate of 5%, the traffic volume (vehicles per day) in the year 2012 will be most nearly:

(A) 40,000
(B) 42,000
(C) 44,000
(D) 46,000

507. It is proposed to construct a left turn lane at a signalized intersection. The percentage of trucks is 4%, and the traffic signal will have a cycle length of 60 sec. Traffic data collected for the left turn volume are shown below.

Time Period	Left Turn Volume
7:00–7:15	15
7:15–7:30	30
7:30–7:45	35
7:45–8:00	25

The minimum storage length (ft) for the proposed left turn lane is most nearly:

(A) 50
(B) 100
(C) 125
(D) 375

GO ON TO THE NEXT PAGE

508. The existing profile for a two-lane highway is shown in the figure below. Because of limited passing sight distance, passing is not allowed. Therefore, to enhance safety, the crest of the curve will be lowered enough to provide minimum passing sight distance. If the design speed is 50 mph, the minimum length L (ft) of vertical curve required is most nearly:

(A) 1,700
(B) 1,800
(C) 2,900
(D) 4,300

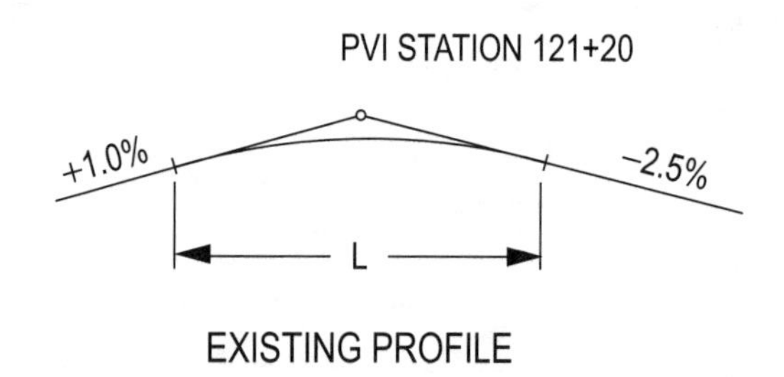

509. A two-lane highway with a design speed of 50 mph and ADT of 4,000 has a 4:1 (H:V) fill side slope. Based on the curves shown below, the suggested horizontal clear zone distance (ft) is most nearly:

(A) 16
(B) 26
(C) 42
(D) 52

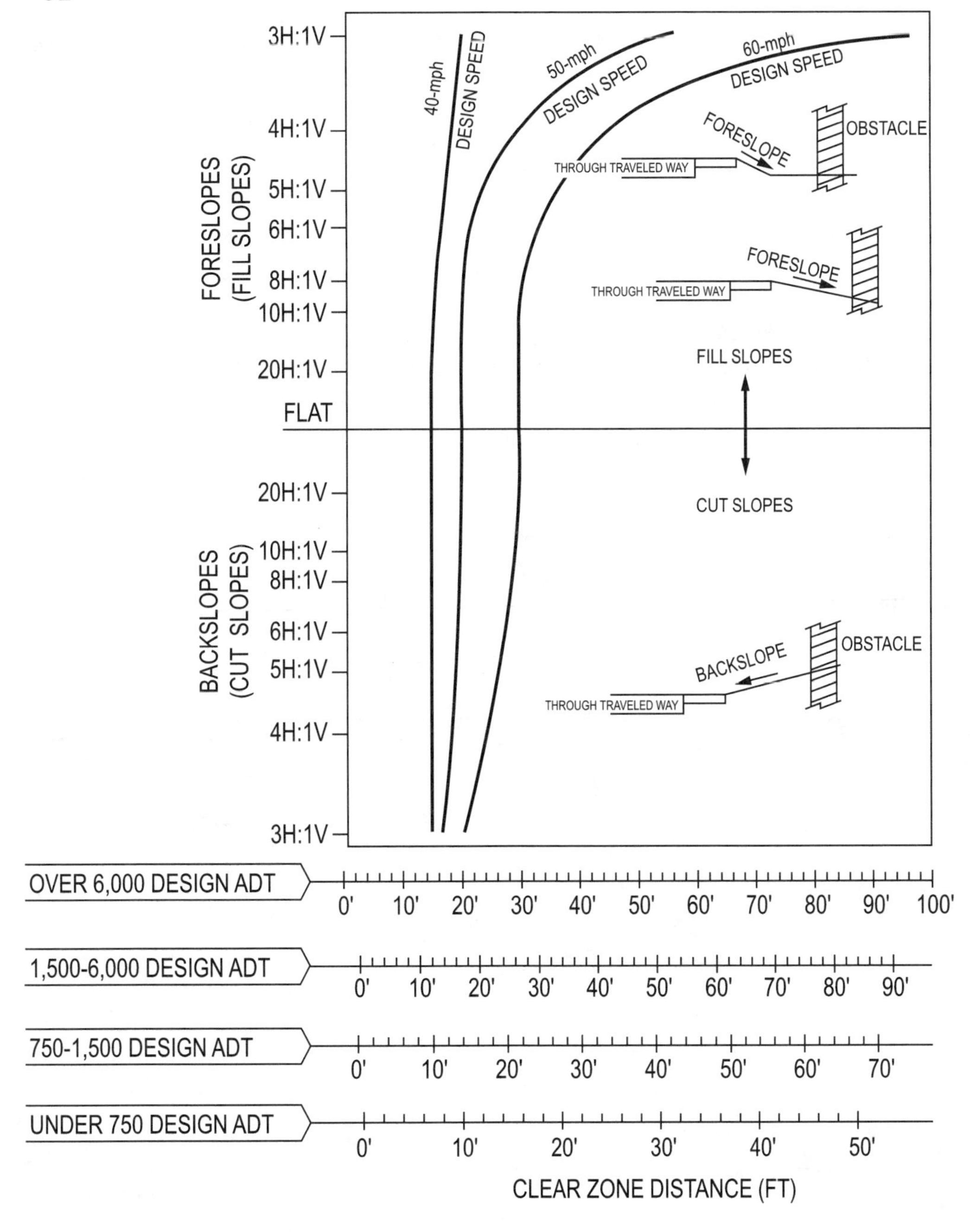

From *Roadside Design Guide*, American Association of State Highway and Transportation Officials, Washington, D.C., 2002. Used by permission.

510. The tangent vertical alignment of a section of proposed highway is shown in the figure below. The station of the high point is most nearly:

(A) 35+00
(B) 42+00
(C) 43+40
(D) 45+15

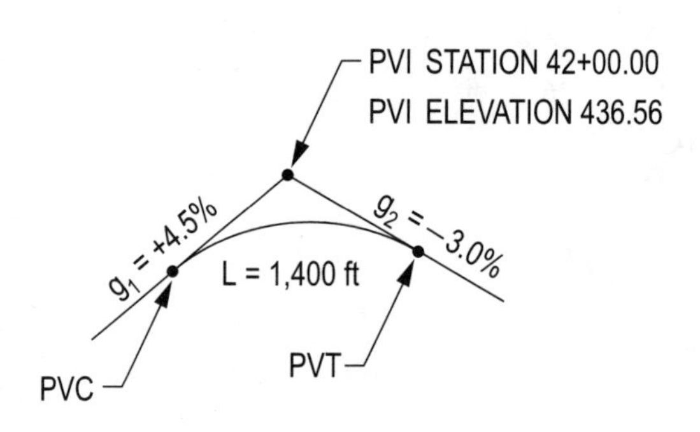

511. A vehicle ran through a stop sign and collided with a tree at an estimated speed of 30 mph. Skid marks were visible beginning at a point 150 ft in advance of and continuing through the intersection, as shown below. Assuming a flat grade, the estimated speed (mph) when braking began was most nearly:

(A) 86
(B) 71
(C) 68
(D) 64

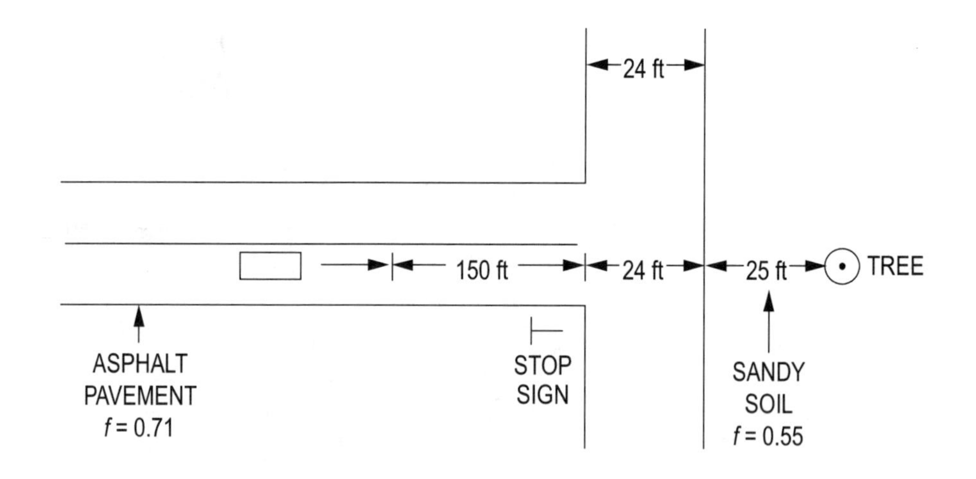

512. A freeway with a speed limit of 55 mph and 12-ft lanes requires a work zone lane shift as shown. The minimum recommended length (ft) for the taper is most nearly:

(A) 220
(B) 305
(C) 330
(D) 660

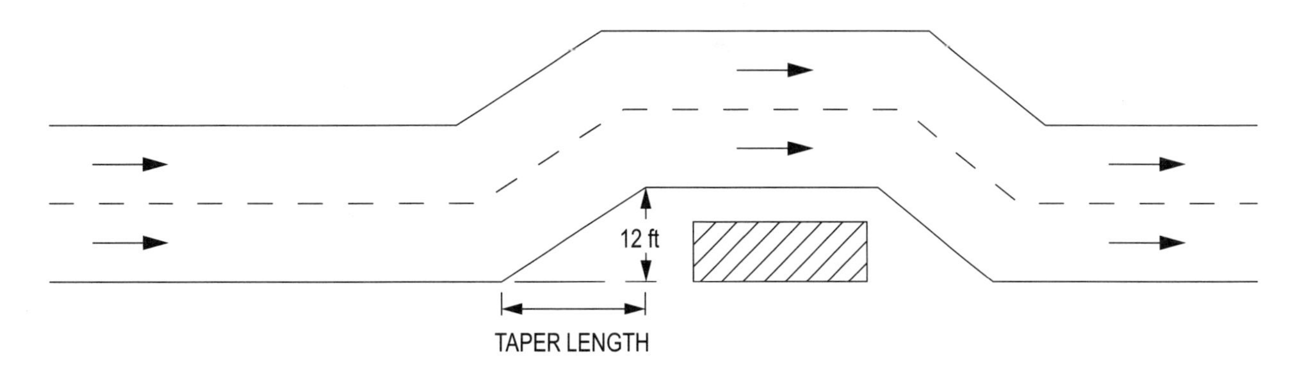

513. A vehicle is approaching a railroad track at a speed of 45 mph. Assume AASHTO-recommended design values for perception-reaction time.

The stop line is located 15 ft from the nearside rail, and the driver is located 5 ft back from the front bumper of the vehicle. The required sight triangle distance (ft) along the highway for a vehicle to stop at the stop line for an approaching train is most nearly:

(A) 295
(B) 347
(C) 380
(D) 438

GO ON TO THE NEXT PAGE

514. A horizontal curve on a two-lane rural highway has the following characteristics:

Design speed, v	60 mph
Radius (minimum safe)	1,091 ft
Coefficient of side friction	0.12
Lane width	12 ft

The rate of superelevation required for this curve is most nearly:

(A) 7%
(B) 10%
(C) 11%
(D) 33%

515. Two highways have tangent horizontal alignments that intersect as shown below. Each highway has a vertical alignment that consists of a vertical curve in the vicinity of the intersection point. Data for the vertical curve on A-2 are as follows:

PVI station	67+00
PVI elevation	103.00 ft
Length of vertical curve	1,000 ft
Tangent grades:	
g_1	–2.0%
g_2	+3.0%

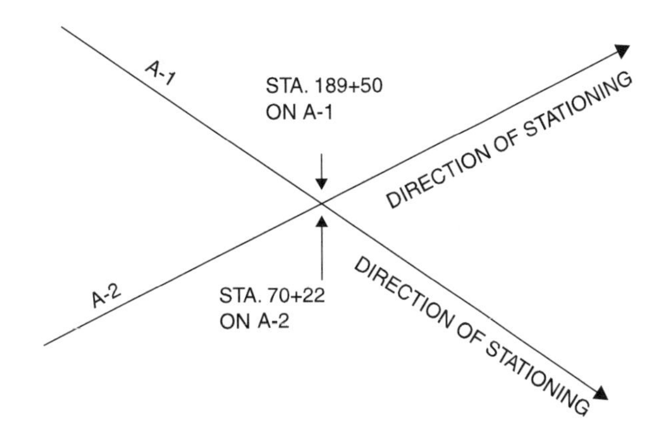

The elevation (ft) on A-2 at the intersection of these two highways is most nearly:

(A) 113.50
(B) 112.00
(C) 109.25
(D) 103.80

516. A single-lane entrance ramp joins a tangent section of freeway mainline as a parallel-type entrance. The entrance ramp design speed is 30 mph, and the highway design speed is 70 mph. The grade is +1.0%. The minimum acceleration length L (ft) needed for the entrance is most nearly:

(A) 110
(B) 520
(C) 1,230
(D) 1,350

517. The swale shown below has a Manning's roughness coefficient of 0.02 and a slope of 0.5%. At a flow of 30 cfs, the depth (in.) of water in the swale is most nearly:

(A) 6
(B) 9
(C) 12
(D) 15

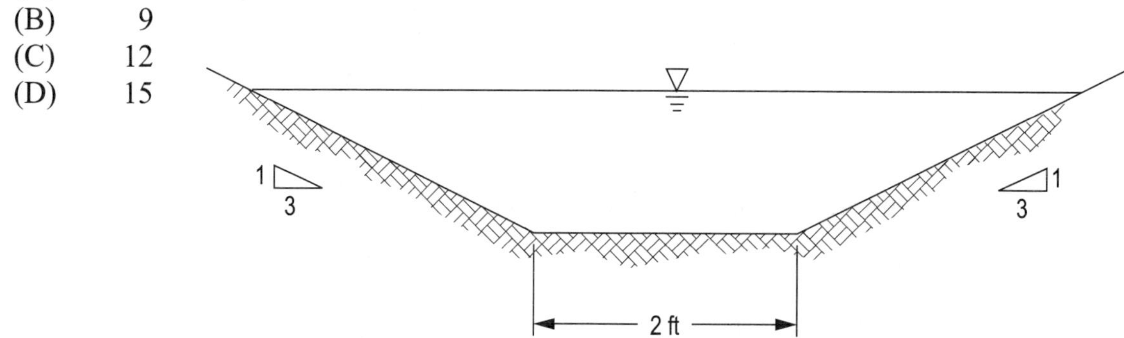

518. A proposed storm sewer will have a slope of 0.20%. The design flow for the line has been determined to be 16 cfs. Assume steady, uniform flow and a Manning's roughness coefficient of 0.012 that is constant for all depths of flow. The minimum circular pipe size (in.) that will accommodate the design flow is most nearly:

(A) 30
(B) 24
(C) 18
(D) 12

519. A parcel of land is to be developed into 1/4-acre-lot single-family housing. Historical rainfall data are as follows:

Time (min)	Rainfall Intensity (in./hr)	
	10-year	100-year
5	2.95	5.10
10	2.08	3.48
20	1.44	2.38
60	0.72	1.30

The parcel drains to a culvert, and the time of concentration is 7.5 min. According to the figure shown, the C value to be used for the 10-year storm is most nearly:

(A) 0.75
(B) 0.79
(C) 0.82
(D) 0.85

519. (Continued)

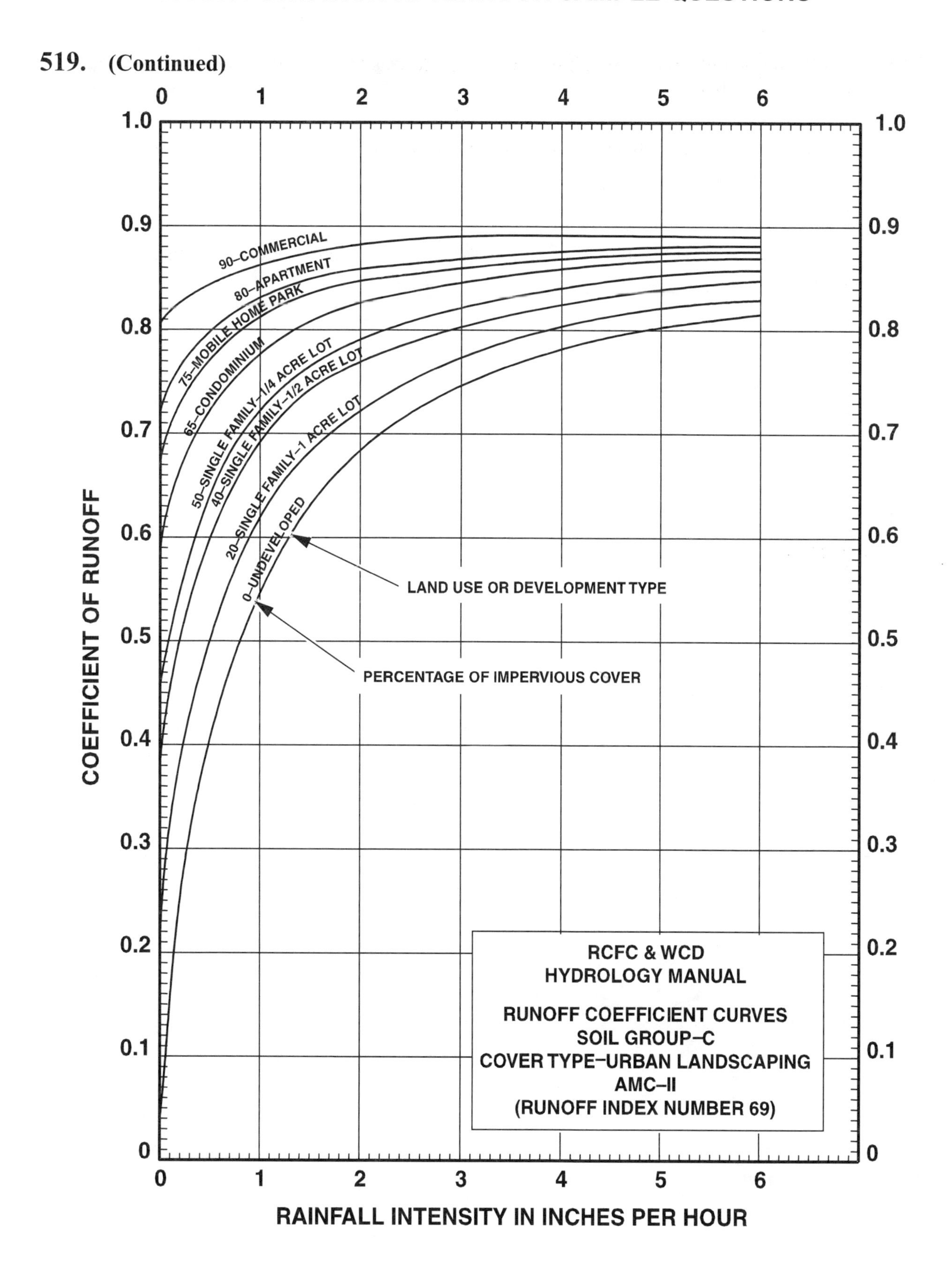

GO ON TO THE NEXT PAGE

520. A soil sample obtained from a highway subgrade was tested in the laboratory with the following results:

Percent passing #200 sieve (based on dry weight)	40%
Liquid limit	56
Plastic limit	47

Using the AASHTO soil classification system, the group index for this soil is most nearly:

(A) 0
(B) 1
(C) 5.5
(D) 11

WATER RESOURCES AND ENVIRONMENTAL AFTERNOON SAMPLE QUESTIONS

This book contains 20 water resources and environmental depth questions, half the number on the actual exam.

501. A 15,000-ft-long, 8-in.-diameter PVC line (C = 140) serves 500 connections at the end of the line. The elevation of the hydraulic grade line is 495 ft at the beginning and 365 ft at the end. The pressure (psi) at the end of the line during peak conditions (1 gpm/connection) is most nearly:

(A) 12
(B) 28
(C) 56
(D) 158

GO ON TO THE NEXT PAGE

502. Your firm is designing a new warehouse. The fire marshal has determined a required fire flow at the site of 3,500 gpm with a residual pressure of 20 psi. You have talked to the city water resources department. The city water system supply is shown in the sketch below. Under the worst demand scenario, the supply is as shown and the demand of the remaining system is as shown. The minimum pipe size (in.) that is required between Point A and the warehouse site to provide the required fire protection is most nearly:

(A) 12
(B) 14
(C) 16
(D) 18

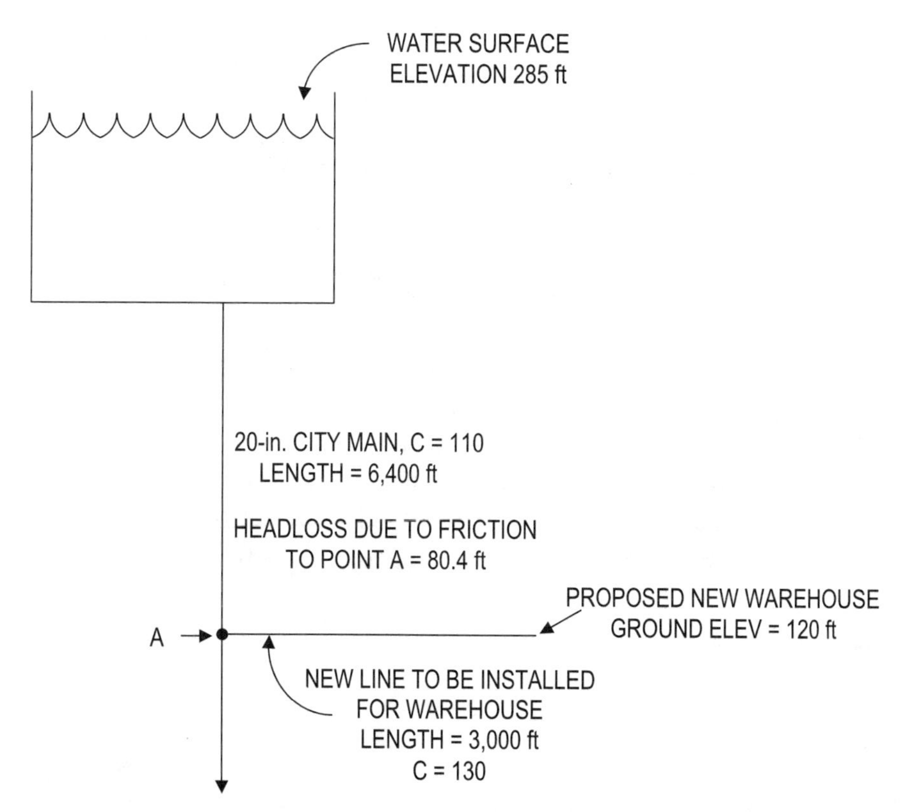

503. A 2,000-ft-long pipeline conveys water at 70°F from a reservoir with a water elevation of 100 ft to a lower reservoir. The pipe is 12-in.-diameter non-riveted steel pipe with a fully open globe valve and a 1/4 closed gate valve. Both the pipe entrance and exit are square edged. Assuming a flow of 7 cfs and a friction factor of 0.015, the elevation of the lower reservoir (ft) is most nearly:

(A) 47.4
(B) 52.6
(C) 56.2
(D) 84.4

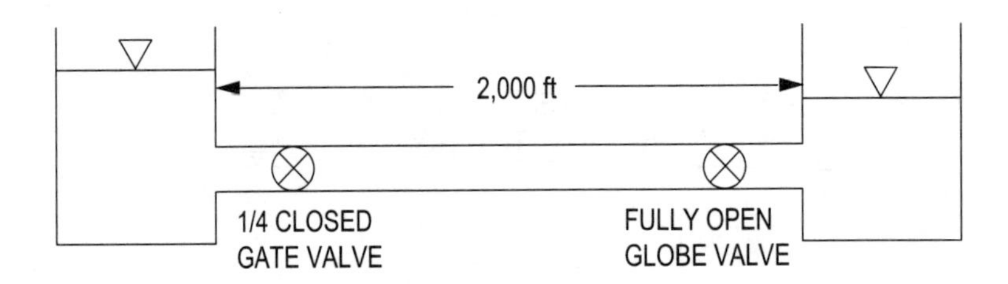

Typical Loss Coefficients		
Gate valve	Fully open	0.19
	1/4 closed	1.15
	1/2 closed	5.6
	3/4 closed	24
Globe valve	Fully open	10
Pipe exit	Square edged	1.0
Pipe entrance	Square edged	0.50

504. The average velocity (fps) of a steady uniform flow in a 15-in.-diameter sewer line with a slope of 0.35%, a depth of 3 in., and a Manning's roughness coefficient of 0.012 is most nearly:

(A) 0.4
(B) 2.1
(C) 3.0
(D) 6.0

505. A long concrete-lined drainage channel has a slope of 0.001 ft/ft, a bottom width of 10 ft and 2:1 (H:V) side slopes. The water depth is 5 ft, and the Manning's roughness coefficient is 0.012. The Froude number for the channel is most nearly:

(A) 0.08
(B) 0.18
(C) 0.66
(D) 0.80

506. Flume and weir formulas require flow rate (Q) in cfs and head (H) in ft. The system below experiences a flow rate of 3.0 MGD. Ignoring the losses in the 24-in. line between the Parshall flume structure and the weir structure, the water surface elevations at Points A and B, respectively, are most nearly:

(A) 100.99, 100.41
(B) 100.99, 100.57
(C) 101.31, 100.69
(D) 101.36, 100.94

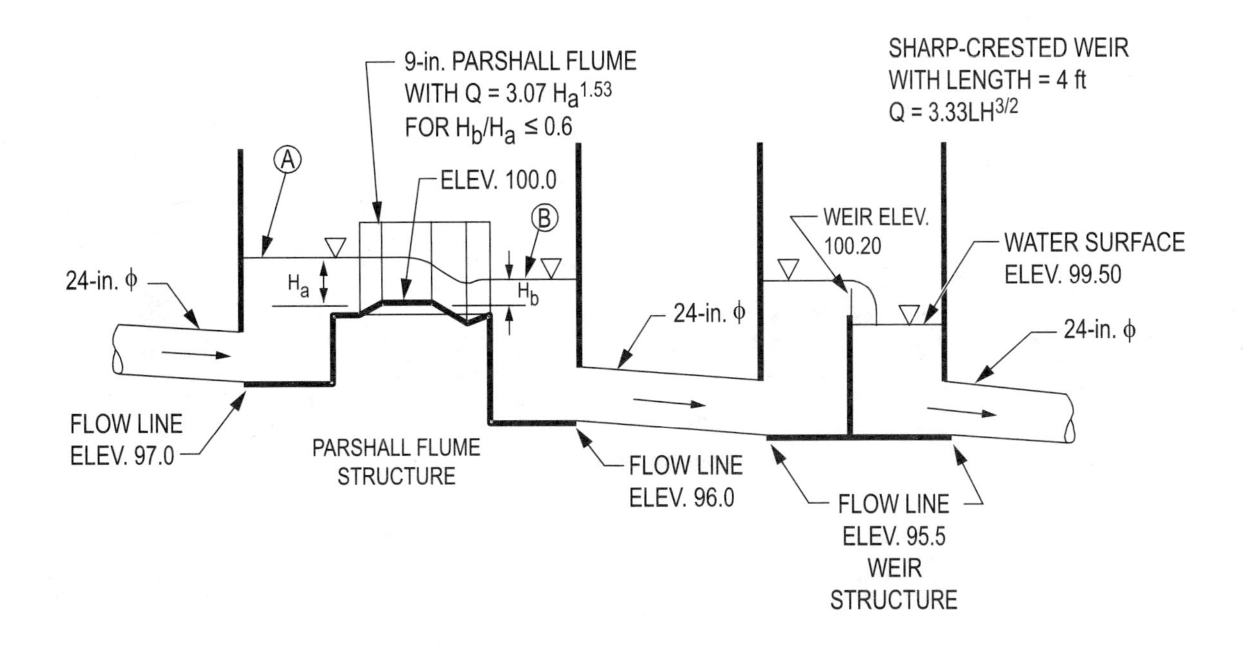

GO ON TO THE NEXT PAGE

507. You are asked to design a culvert for a highway. If you want to be 90% confident that the capacity of the culvert will not be exceeded in a 30-year period, the return period storm (years) you should use in the design should be most nearly:

(A) 1
(B) 14
(C) 30
(D) 285

508. A rain event has an intensity of 1.5 in./hr for the first hour followed by 0.7 in./hr for the second hour. The 1-hour unit hydrograph for the watershed is given as:

Q (cfs/in.)	0.5	1.2	0.4
T (hours)	1	2	3

Neglecting infiltration, the discharge (cfs) from the watershed during the second hour is most nearly:

(A) 2.64
(B) 2.15
(C) 1.44
(D) 1.20

GO ON TO THE NEXT PAGE

509. Assume that you are evaluating an agricultural watershed. The soil is classified as clay with a high swelling potential. The watershed is a pasture that has 65% ground cover and is not heavily grazed. The potential maximum retention (in.) after runoff begins (also called the soil storage capacity) is most nearly:

(A) 1.24
(B) 1.83
(C) 2.65
(D) 4.49

510. Regional groundwater flow through a sandstone aquifer is shown in the figure below. The aquifer has an average thickness of 650 ft. The distance from the recharge area to the discharge point is 12 miles, and the head difference is 200 ft. The hydraulic conductivity is 15 ft/day. The time (years) it will take for water to travel from the recharge area to the discharge area is most nearly:

(A) 0.5
(B) 20
(C) 900
(D) 3,700

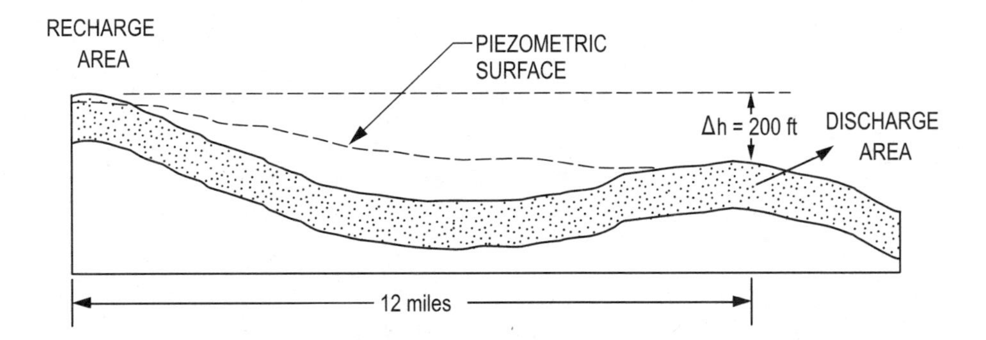

511. A secondary clarifier is being designed for a 3.5-MGD wastewater treatment plant. The mixed liquor suspended solids (MLSS) entering the clarifier from the aeration tank is 2,500 mg/L SS, and the return activated sludge is 7,790 mg/L SS. Assuming 70,000 gal of waste activated sludge at 7,790 mg/L is drawn off daily and negligible solids in the clarifier effluent, the design return activated sludge flow rate (MGD) is most nearly:

(A) 3.40
(B) 1.55
(C) 1.00
(D) 0.75

512. The fully nitrified influent to the anoxic basin contains 25 mg/L nitrate-N. The following equation applies:

$$NO_3^- + 1.08\,CH_3OH + H^+ \rightarrow 0.065\,C_5H_7O_2N + 0.47\,N_2 + 0.76\,CO_2 + 2.44\,H_2O$$

The minimum methanol requirement (mg/L) to provide complete denitrification is most nearly:

(A) 14
(B) 23
(C) 57
(D) 62

GO ON TO THE NEXT PAGE

513. A 5-MGD municipal wastewater treatment plant uses a two-stage trickling filter process composed of two parallel treatment trains. The BOD_5 concentration of the treatment plant influent is 200 mg/L, and the first-stage BOD_5 removal efficiency is 70%. The recirculation factor F is 2. Using the NRC equation method, the calculated volume (ft^3) of each of the first-stage filters is most nearly:

(A) 40
(B) 1,200
(C) 36,000
(D) 71,000

514. Industry X discharges 5 MGD of effluent into a river at Point P. The average background flow rate of the river is 100 cfs. The plot below shows the in-stream DO profile downstream of P.

Industry Y will be allowed to discharge 5 MGD at Point Q, 30 miles downstream of P, provided the in-stream DO at Q does not fall below 4.25 mg/L after mixing with the effluent from Industry Y. The DO in the effluent of Industry Y is 3 mg/L. To meet these conditions, Industry Y decides to add DO to their effluent before discharging into the river. The amount of oxygen (lb/day) to be added would be most nearly:

(A) 321
(B) 198
(C) 177
(D) 52

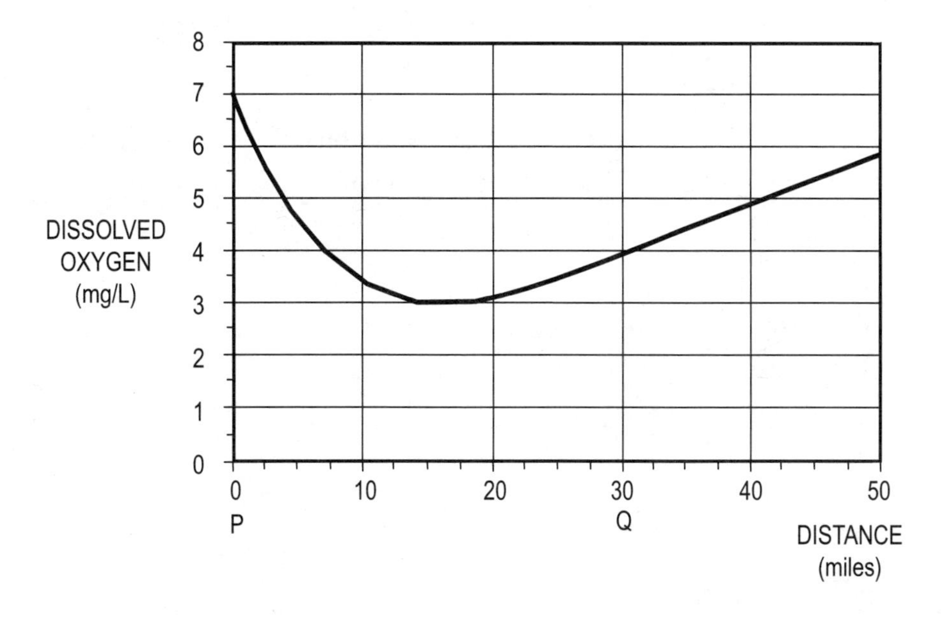

GO ON TO THE NEXT PAGE

515. Selected water quality parameters for a river and a tributary are as follows:

Parameter	River	Tributary
Flow rate (cfs)	50	20
Total phosphorus (mg/L)	0.02	0.5
Chlorophyll (mg/L)	0.05	0.1

Based on the above data and recognizing that algae growth in this system is limited by phosphorus concentrations less than 0.015 mg/L, the longitudinal profile of the algae concentration in the main river downstream of the confluence point will probably:

(A) stay the same

(B) decrease continuously

(C) decrease first and then increase

(D) increase first and then decrease

516. An industry proposes to discharge its effluent into a nearby stream. Average flow rates and temperature of the effluent and the river are as follows:

Parameter	Effluent	River
Flow rate	10 MGD	100 cfs
Temperature	42°C	19°C

From a BOD test under laboratory conditions, the rate constant k_1 for the effluent was found to be 0.1 day^{-1} at 20°C. The in-stream value of k_1 (day^{-1}) would be most nearly:

(A) 0.08
(B) 0.11
(C) 0.17
(D) 0.33

517. A multipurpose reservoir has been designed with 22% of its volume for sediment storage. The reservoir's initial volume is 1.41×10^{10} ft^3. The influent streams to the reservoir have a total flow rate of 2.1×10^9 ft^3/day with an average suspended solid concentration of 172 mg/L. Assume the average sediment density is 80.0 pcf, and the outflow from the reservoir has negligible suspended solids. The expected lifetime (years) of the reservoir is most nearly:

(A) 3
(B) 30
(C) 57
(D) 110

518. A community is served by a well that pumps 350 gpm through 1,800 ft of 8-in. pipe prior to the first customer. Chlorine is injected into the pipe at the well site. The water system must provide 99.99% removal/inactivation of viruses. If the water pH is 7.5 and the temperature is 15°C, the minimum free chlorine residual (mg/L) required at the first customer is most nearly:

(A) 0.3
(B) 0.5
(C) 1.0
(D) 2.3

C · t* Values for Inactivation of Viruses at Various Temperatures and pH 6–9						
	WATER TEMPERATURE					
Parameter	**LOG INACTI-VATION**	**0.5°C [(mg/l)·min]**	**5°C [(mg/l)·min]**	**10°C [(mg/l)·min]**	**15°C [(mg/l)·min]**	**20°C [(mg/l)·min]**
Free chlorine	2.0	6	4	3	2	1
	3.0	9	6	4	3	2
	4.0	12	8	6	4	3
Preformed chloramine	2.0	1,200	860	640	430	320
	3.0	2,100	1,400	1,100	710	530
Chlorine dioxide	2.0	8.4	5.6	4.2	2.8	2.1
	3.0	25.6	17.1	12.8	8.6	6.4
Ozone	2.0	0.9	0.6	0.5	0.3	0.2
	3.0	1.4	0.9	0.8	0.5	0.4

Source: Adapted from *Guidance Manual for Compliance with the Filtration and Disinfection Requirements for Public Water Systems Using Surface Water Sources* (Environmental Protection Agency, 1991).

* *C · t* values for free chlorine, ozone, and chlorine dioxide include safety (uncertainty) factors. The chloramine values are based on laboratory data using preformed chloramine to inactivate hepatitis A and do not include a safety factor.

GO ON TO THE NEXT PAGE

519. The constituents of a water supply are shown in the bar chart. There is no magnesium. It is desired to remove the carbonate hardness due to calcium. No noncarbonate hardness will be removed. Excess lime beyond the stoichiometric amount will be added at a dose of 20 mg/L as $CaCO_3$. The total amount of lime (mg/L as $CaCO_3$) that must be added is most nearly:

(A) 140
(B) 150
(C) 190
(D) 220

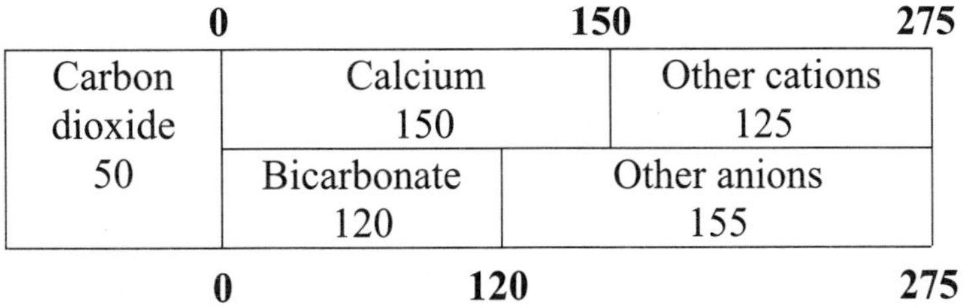

	0	150	275
Carbon dioxide 50	Calcium 150	Other cations 125	
	Bicarbonate 120	Other anions 155	
	0	120	275

Length of bars not to scale.
All concentrations are mg/L as $CaCO_3$.

520. A utility company can borrow at 6% interest. The company can buy a packaged wastewater plant for $10,000,000 payable up-front and estimates $200,000/year for maintenance. Another option is to lease the plant with a $1,000,000 down payment and $600,000 semiannual payments (including maintenance) for the 20-year lease period. Assume the plant has a useful life of 20 years and a salvage value of zero. Which of the following statements is most accurate?

(A) Buying is better by $2.6 million.

(B) Buying is better by $4.9 million.

(C) Leasing is better by $4.4 million.

(D) Buying and leasing costs are about the same.

CIVIL BREADTH
MORNING SOLUTIONS

Correct Answers to the MORNING Sample Questions

Detailed solutions for each question begin on the next page.

101	C
102	D
103	D
104	A
105	B
106	D
107	C
108	B
109	C
110	D
111	D
112	A
113	B
114	D
115	B
116	C
117	A
118	A
119	B
120	A

MORNING SOLUTIONS

101. Use Average End Area Method.

Stationing	Excavation (yd^3)	Embankment (yd^3)
1+00 to 2+00	$\frac{50+150}{2}\times\frac{100}{27}=370$	
2+00 to 3+00	$\frac{50+0}{2}\times\frac{100}{27}=93$	$\frac{0+40}{2}\times\frac{100}{27}=74$
Total	**463**	**74**

Net excavation = $463-74=389\ yd^3$

THE CORRECT ANSWER IS: (C)

102. Reference: Peurifoy and Oberlender, *Estimating Construction Costs,* 5th ed., Chapter 1, p. 11, Quantity Takeoff.

$$\text{Horizontal length of side slope} = \frac{14}{2}\times 3 = 21.0\text{ ft}$$

$$\text{Slope length} = \sqrt{(14)^2+(21)^2} = 25.24\text{ ft}$$

$$\text{Cross-sectional area of lining} = \left[(2\times 25.24)+9\right]\frac{7}{12} = 34.70\text{ ft}^2$$

$$\text{Volume of lining} = \frac{(34.70\times 227)}{27} = 291.7\text{ yd}^3$$

$$\text{Delivered volume} = 291.7\text{ yd}^3\times \underset{\text{(waste)}}{1.12} = 327\text{ yd}^3$$

THE CORRECT ANSWER IS: (D)

MORNING SOLUTIONS

103. Reference: Callahan, Quackenbush, and Rowings, *Construction Project Scheduling*, 1992.

$TF = LF - EF$

$EF = ES + D$

$TF = LF - ES - D$

For Activity B:

$TF_B = LF_B - ES_B - D_B$

THE CORRECT ANSWER IS: (D)

104. Reference: Portland Cement Association, *Design and Control of Concrete Mixtures.*

Water tightness is the ability of concrete to hold or retain water without visible leakage. Generally, less permeable concrete is more watertight. A lower water-cement ratio reduces permeability, thereby increasing water tightness.

THE CORRECT ANSWER IS: (A)

105. The depth below the original ground surface to the center of the clay layer is 15 ft (5 ft of moist sand, 5 ft of saturated sand, and 5 ft of clay), which is 10 ft below the water table.

σ'_o = weight of soil above a depth of 15 ft minus the weight of water below the water table

$= \Sigma \gamma \Delta h - \gamma_w h_w$

$= (115)(5.0) + (130)(5.0) + (95)(5.0) - (62.4)(10.0) = 1{,}700 - 624$

$= 1{,}076$ psf

THE CORRECT ANSWER IS: (B)

106. This is a fine-grained soil since more than 50% passes the #200 sieve. The liquid limit and plasticity index plots above the A line and is greater than 50, so soil is a CH.

MAJOR DIVISIONS			GROUP SYMBOLS	TYPICAL NAMES	CLASSIFICATION CRITERIA		
COARSE-GRAINED SOILS More than 50% retained on 0.075 mm (No. 200) sieve	GRAVELS 50% or more of coarse fraction retained on 4.75 mm (No. 4) sieve	CLEAN GRAVELS	GW	Well-graded gravels and gravel-sand mixtures, little or no fines	Classification on basis of percentage of fines Less than 5% Pass 0.075 mm sieve GW, GP, SW, SP More than 12% Pass 0.075 mm sieve GM, GC, SM, SC 5% to 12% Pass 0.075 mm sieve Border Classification requiring use of dual symbols	$C_u = D_{60}/D_{10}$ Greater than 4 $C_z = \frac{(D_{30})^2}{D_{10} \times D_{60}}$ Between 1 and 3	
			GP	Poorly graded gravels and gravel-sand mixtures, little or no fines		Not meeting both criteria for GW	
		GRAVELS WITH FINES	GM	Silty gravels, gravel-sand-silt mixtures		Atterberg limits plot below "A" line or plasticity index less than 4	Atterberg limits plotting in hatched area are borderline classifications requiring use of dual symbols
			GC	Clayey gravels, gravel-sand-clay mixtures		Atterberg limits plot above "A" line and plasticity index greater than 7	
	SANDS More than 50% of coarse fraction passes 4.75 mm (No. 4) sieve	CLEAN SANDS	SW	Well-graded sands and gravelly sands, little or no fines		$C_u = D_{60}/D_{10}$ Greater than 6 $C_z = \frac{(D_{30})^2}{D_{10} \times D_{60}}$ Between 1 and 3	
			SP	Poorly graded sands and gravelly sands, little or no fines		Not meeting both criteria for SW	
		SANDS WITH FINES	SM	Silty sands, sand-silt mixtures		Atterberg limits plot below "A" line or plasticity index less than 4	Atterberg limits plotting in hatched area are borderline classifications requiring use of dual symbols
			SC	Clayey sands, sand-clay mixtures		Atterberg limits plot above "A" line and plasticity index greater than 7	
FINE-GRAINED SOILS 50% or more passing 0.075 mm (No. 200) sieve	SILTS AND CLAYS	Liquid limit 50% or less	ML	Inorganic silts, very fine sands, rock flour, silty or clayey fine sands	PLASTICITY CHART (see below)		
			CL	Inorganic clays of low to medium plasticity, gravelly clays, sandy clays, silty clays, lean clays			
			OL	Organic silts and organic silty clays of low plasticity			
	SILTS AND CLAYS	Liquid limit 50% or greater	MH	Inorganic silts, micaceous or diatomaceous fine sands or silts, elastic silts			
			CH	Inorganic clays of high plasticity, fat clays			
			OH	Organic clays of medium to high plasticity			
Highly Organic Soils			PT	Peat, muck and other highly organic soils	Visual-manual identification, See ASTM Designation D 2488.		

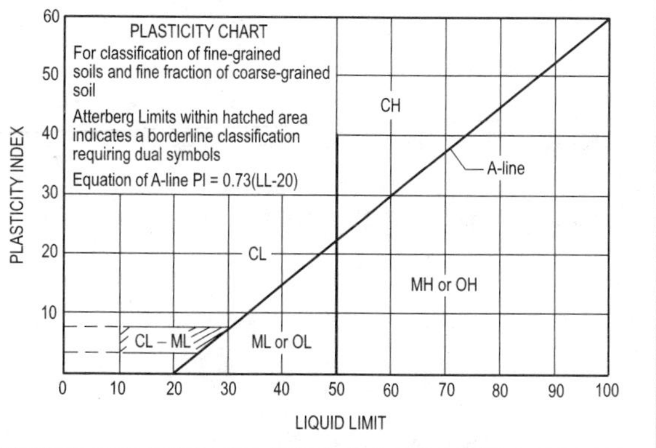

PLASTICITY CHART
For classification of fine-grained soils and fine fraction of coarse-grained soil
Atterberg Limits within hatched area indicates a borderline classification requiring dual symbols
Equation of A-line PI = 0.73(LL-20)
PLASTICITY INDEX
LIQUID LIMIT
CH
A-line
CL
MH or OH
CL – ML
ML or OL
0 10 20 30 40 50 60 70 80 90 100
10 20 30 40 50 60

The above material is adapted, with permission, from ASTM D2488-06, "Standard Practice for Description and Identification of Soils (Visual-Manual Procedure)," copyright ASTM International, 100 Barr Harbor Drive, West Conshohocken, PA 19428.

THE CORRECT ANSWER IS: (D)

107. The solution must be based on dry, not moist, unit weights.

The weight of water to be added is

$$W_w = (500{,}000\ \text{yd}^3)\ (27\ \text{ft}^3/\text{yd}^3)\ (0.90)\ (116.0\ \text{pcf})\ (0.05) = 70.47 \times 10^6\ \text{lb}$$

The volume of water added is

$$V_w = (70.47 \times 10^6\ \text{lb})/(8.33\ \text{lb/gal})$$
$$= 8.46 \times 10^6\ \text{gal}$$

THE CORRECT ANSWER IS: (C)

108. $K_A = \tan^2\left(45 - \frac{\phi}{2}\right) = \tan^2\left(45 - \frac{28}{2}\right)$

$K_A = 0.36$

THE CORRECT ANSWER IS: (B)

109. Dead load plus wind load gives the maximum uplift.

THE CORRECT ANSWER IS: (C)

110. Sum moments about Point B.

$$(R_A)(32\ \text{ft}) - [0.45\ \text{kips/ft} \times (3\ \text{ft} + 29\ \text{ft}) \times 6\ \text{ft}] - \left(\frac{0.45\ \text{kips/ft}}{\cos 39.5°} \times 20\ \text{ft} \times 16\ \text{ft}\right) - (1\ \text{kip} \times 25\ \text{ft})$$
$$- (1\ \text{kip} \times 16.5\ \text{ft}) = 0$$

$R_A = 9.8\ \text{kips}$

THE CORRECT ANSWER IS: (D)

MORNING SOLUTIONS

111. Deflection will not change. F_y is not part of deflection calculations.

THE CORRECT ANSWER IS: (D)

112. Tube sections are better able to resist lateral torsional buckling than channel, wide-flange, and double-angle sections and are therefore generally more efficient for use as beams with long unbraced lengths.

THE CORRECT ANSWER IS: (A)

113. Compute the tangent elevation at Station 73+00.

Tangent elevation = 334.56 + (3.0)(3.0) = 343.56 ft

Compute the tangent offset, e, at the PVI station.

$$e = (L/8)(g_2 - g_1) = (15/8)\ [+2.0\% - (-3.0\%)] = 9.375 \text{ ft}$$

Compute the tangent offset, y, at Station 73+00.

$$y = (4e / L^2)\ X^2 = [(4)(9.375)(4.5)^2]/(15)(15) = 3.38 \text{ ft}$$

Compute the vertical curve elevation at Station 73+00.

Curve elevation = tangent elevation + tangent offset

= 343.56 + 3.38 = 346.94 ft

Compute the clearance between the bridge and the vertical curve at Station 73+00.

Clearance = bridge elevation – curve elevation

= 365.94 – 346.94 = 19.0 ft

THE CORRECT ANSWER IS: (B)

114. Reference: *Highway Capacity Manual,* 1998, pp. 1–4.
There are six levels of service: A through F.

THE CORRECT ANSWER IS: (D)

115. Referring to the figure below, determine T, the distance from the PC to the PI.

$$T = R \tan\frac{\Delta}{2} = (2{,}550 \text{ ft})\left(\tan\frac{78^\circ 35'30''}{2}\right) = 2{,}086.84 \text{ ft}$$

Determine the station of the PI.

$$\begin{aligned} PI &= PC + 2{,}086.84 \\ &= 12{+}56.00 + 2{,}086.84 \\ &= 33{+}42.84 \end{aligned}$$

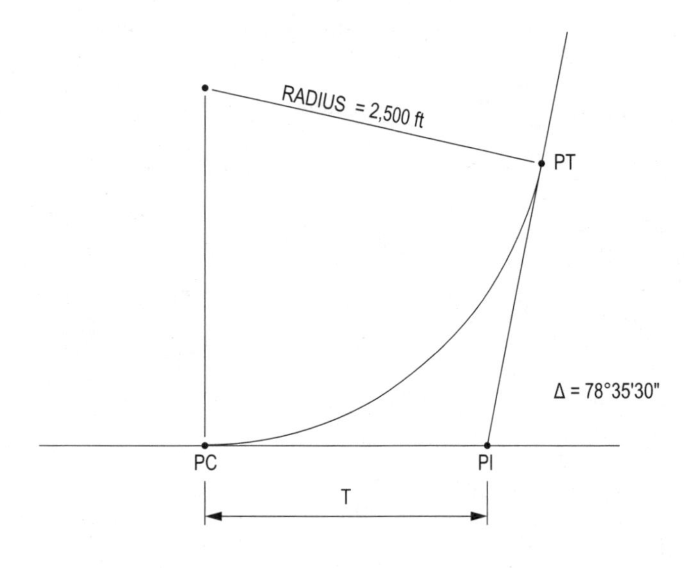

THE CORRECT ANSWER IS: (B)

116. Stations on vertical curves are based on horizontal distances. Therefore, the required horizontal distance can be computed as follows:

Horizontal distance = (76+00.00) – (42+00.00) = 34+00.00
= 3,400 ft

THE CORRECT ANSWER IS: (C)

117. Reference: Mays, *Hydraulic Design Handbook,* pp. 10.16–10.19.

Must select a pump speed (typically between 500–1,800 rpm). Calculate specific speed, N_S.

$$N_s = \frac{\text{rpm}(Q)^{1/2}}{H^{3/4}} = \frac{(1,800)\left[10\text{ cfs}\left(7.48\text{ gal/ft}^3\right)\left(60\text{ sec/min}\right)\right]^{1/2}}{(1,200\text{ ft})^{3/4}} = 591$$

Select a radial flow impeller.

THE CORRECT ANSWER IS: (A)

118. Reference: Chow, *Open-Channel Hydraulics*, 4th ed., p. 393 ff.

A hydraulic jump occurs when the flow in an open channel goes from super-critical to subcritical. The flow velocity decreases and the depth increases. (A) is the most common cause of a hydraulic jump.

THE CORRECT ANSWER IS: (A)

119. Reference: Chow, *Handbook of Applied Hydrology,* 1964, pp. 14–17.

Time	Q_u	Q_1 +	Q_2 =	Q_{total}
0	0	0	0	0
1	75	37.5	0	37.5
2	200	100	37.5	137.5
3	100	50	100	150
4	50	25	50	75
5	25	12.5	25	37.5
6	0	0	12.5	12.5
				450 cfs

$$450 \frac{\text{ft}^3}{\text{sec}} \times \frac{3{,}600 \text{ sec}}{\text{hr}} \times 1 \text{ hr} \times \frac{1 \text{ acre}}{43{,}560 \text{ ft}^2} = 37.19 \text{ acre-ft}$$

THE CORRECT ANSWER IS: (B)

120. Reference: Peavy, Rowe, and Tchobanoglous, *Environmental Engineering,* 1985, pp. 355–359.

The minimum velocity provides scour, which prevents the deposition of solids.

THE CORRECT ANSWER IS: (A)

CONSTRUCTION
AFTERNOON SOLUTIONS

Correct Answers to the CONSTRUCTION Afternoon Sample Questions

Detailed solutions for each question begin on the next page.

501	B
502	A
503	B
504	B
505	C
506	D
507	D
508	A
509	C
510	D
511	B
512	D
513	D
514	C
515	B
516	C
517	A
518	C
519	C
520	B

CONSTRUCTION AFTERNOON SOLUTIONS

501. References: Peurifoy and Schexnayder, *Construction Planning, Equipment and Materials,* 6th ed., Chapter 10; and Nunnally, *Construction Methods and Management,* 2nd ed., Chapters 2, 3, and 4.

Density of embankment fill, $\gamma_{dry} = (0.90)(120.0 \text{ pcf}) = 108.0 \text{ pcf}$

The total weight of dry soil required is:

$$W_{total} = (500{,}000 \text{ yd}^3)(27 \text{ ft}^3/\text{yd}^3)(108.0 \text{ pcf}) = 1.458 \times 10^9 \text{ lb}$$

The dry unit weight of soil in the truck is:

$$\gamma_{dry} = G_s\, \gamma_w/(1 + e) = (2.65)(62.4 \text{ pcf})/(1 + 1.30) = 71.9 \text{ pcf}$$

Truck capacity:

$$W_{truck} = (5.0 \text{ yd}^3)(27 \text{ ft}^3/\text{yd}^3)(71.9 \text{ pcf}) = 9{,}700 \text{ lb/truck}$$

Therefore, the minimum number of trucks required is:

$$N = W_{total}/W_{truck} = 1.458 \times 10^9 / 9{,}700 = 150{,}000 \text{ trucks}$$

THE CORRECT ANSWER IS: (B)

502. Reference: Kavanagh, *Surveying with Construction Application*, 2nd ed., Section 12.3, p. 308.

$$R = 2 \text{ Stations} = 200 \text{ ft}$$

$$M = R\left(1 - \cos\frac{\Delta}{2}\right)$$

$$12.8 \text{ ft} = 200 \text{ ft}\left(1 - \cos\frac{\Delta}{2}\right)$$

$$\Delta = 41.22°$$

$$L = 2\pi R \frac{\Delta}{360°}$$

$$L = 2\pi(200 \text{ ft})\left(\frac{41.22°}{360°}\right)$$

$$L = 143.88 \text{ ft}$$

THE CORRECT ANSWER IS: (A)

503. Reference: Frank R. Walker Company, *Walker's Building Estimator's Reference Book*, 28th ed., pp. 1218–1226.

$$\text{Perimeter} = (2 \times 200 \text{ ft}) + (2 \times 100 \text{ ft}) = 600 \text{ ft}$$

$$\text{Wall surface area} = 600 \text{ ft} \times 7 \text{ ft} = 4{,}200 \text{ ft}^2$$

$$\text{No. of gallons} = \frac{4{,}200 \text{ ft}^2/\text{coat}}{300 \text{ ft}^2/\text{gal}} \times 2 \text{ coats} = 28 \text{ gal}$$

THE CORRECT ANSWER IS: (B)

504. References: Collier and Ledbetter, *Engineering Cost Analysis,* 1982, pp. 275–307; and Gerald Smith, *Engineering Economy,* 4th ed., pp. 236–252.

Benefit = reduction in annual costs

= 250,000 – 248,000 = $2,000

$$\text{Costs} = 9{,}000\left(\frac{4}{p}\right)_{5\text{ yr}}^{10\%} + 1{,}000(0.10)$$

= 2,374 + 100 = 2,474

Alternatively,

$$\text{Costs} = 10{,}000\left(\frac{4}{p}\right)_{5\text{ yr}}^{10\%} - 1{,}000\left(\frac{4}{p}\right)_{5\text{ yr}}^{10\%}$$

$$= 2{,}638 - 164 = 2{,}474$$

$$B/C = \frac{2{,}000}{2{,}474} = 0.81$$

THE CORRECT ANSWER IS: (B)

505. Reference: Peurifoy and Oberlender, *Estimating Construction Costs,* 5th ed., Chapter 10.

Erect and strip crew cost:

$$\left[(4\times\$32.73)+(2\times\$26.08)+\$35.37\right]/7\text{ workers}=\frac{\$218.45}{7}=\$31.21/\text{LH}$$

Place concrete crew cost:

$$\left[(3\times\$26.08)+\$32.73+\$35.37\right]/5\text{ workers}=\frac{\$146.34}{5}=\$29.27/\text{LH}$$

Erect forms: $72\times12\times2=1,728$ ft^2@ 5.5 ft^2/LH = 314.18 LH @ \$31.21/LH = \$9,806

Strip forms: 1,728 ft^2@ 15 ft^2/LH = 115.2 LH @ \$31.21/LH = \$3,595

Place concrete: $(72\times12\times1.0)/27=32$ yd^3@ 2.2 yd^3/LH = 14.54 LH@\$29.27/LH = \$426

Forms: [72 (1/3) × 12 × 2 = 576 ft^2@ \$2.66/ft^2] + [72(2/3) × 12 × 2 = 1,152 ft^2@ \$0.34/ft^2] = \$1,924

Concrete = 32 yd^3 × 1.10 = 35 yd^3@ \$97.20/yd^3 = \$3,402

Reinforcing = 32 yd^3 × \$120.00/yd^3 = \$3,840

Total cost = \$9,806 + \$3,595 + \$426 + \$1,924 + \$3,402 + \$3,840 = \$22,993

THE CORRECT ANSWER IS: (C)

506. Reference: Meriam and Kraige, *Engineering Mechanics,* Vol. 1, 3rd ed., Chapter 3.

The center of gravity of the load is offset. The load is therefore heavier on the left side of the rigging. The portion of the 60-kip load on the left side is (40 ft/60 ft)(60 kips) = 40 kips. The load is further amplified by the slope of the sling. The vector length of the sling is $\sqrt{50^2+20^2}=53.85\text{ ft}$. The force in the sling is (53.85 ft/50.00 ft)(40 kips) = 43.1 kips.

THE CORRECT ANSWER IS: (D)

507. References: Peurifoy and Schexnayder, *Construction Planning, Equipment and Materials,* 6th ed., Chapter 10; and Nunnally, *Construction Methods and Management,* 2nd ed., Chapter 4.

This operation is weight limited.

Bank density = 110 pcf $\times$ 27 ft^3/yd^3 = 2,970 lb/yd^3

Truck capacity (by weight) = $\dfrac{27,000}{2,970} = 9.09$ yd^3

Trips per 55 min/hr = $\dfrac{55}{17} = 3.24$

Haulage rate = 3.24 trips/hr $\times$ 9.09 yd^3/trip = 29.45 yd^3/hr

THE CORRECT ANSWER IS: (D)

508. Reference: Oberlender, *Project Management for Engineering and Construction,* 2nd ed., Chapter 8, pp. 139–184.

Network calculations are shown below. Since FF in E $\geq$ 2 days, delay will not affect any other activity.

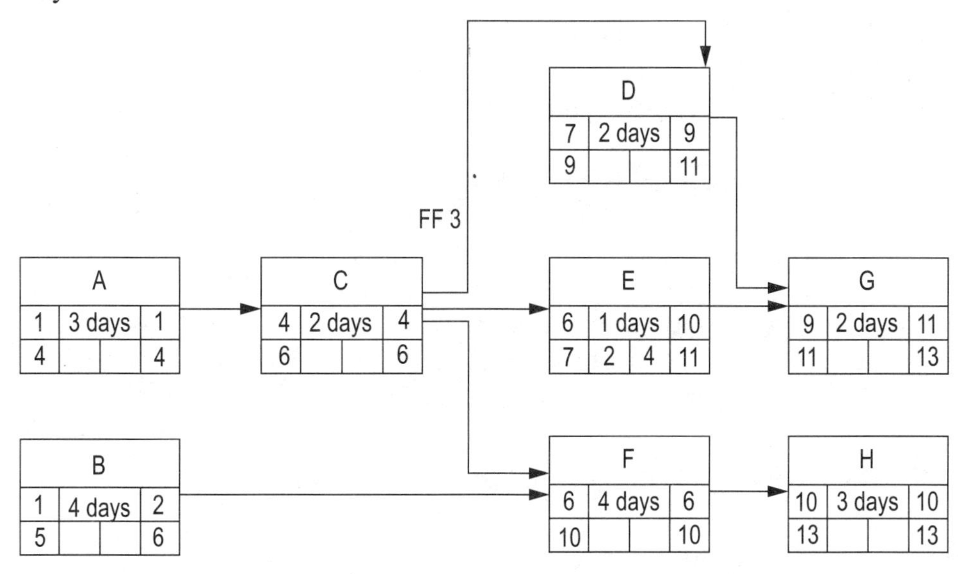

THE CORRECT ANSWER IS: (A)

509. References: Oberlender, *Project Management for Engineering and Construction,* 2nd ed., Chapter 8; and Associated General Contractors of America, *Construction Planning and Scheduling*, 1997, Chapter 6.

The longest path to the start of Activity N is defined by Path C-G-J-L, and the duration along that path is 23. The early start of Activity N is 23.

THE CORRECT ANSWER IS: (C)

510. Reference: Schexnayder and Mayo, *Construction Management Fundamentals,* 2003.

B is the turning point where the job goes from excavation to fill operation. D is the point where it goes back to excavation. Therefore, B–D is the fill operation (Statement II), and D is a transition point (Statement III). Statements II and III are true.

THE CORRECT ANSWER IS: (D)

511. Reference: Oberlender, *Project Management for Engineering and Construction,* 2nd ed., pp. 217–226.

BCWS = \$85,000 planned costs
ACWP = \$95,000 actual spent
BCWP = \$70,000 earned value

Schedule variance = BCWP – BCWS
= \$70,000 – \$85,000
= –\$15,000 $\therefore$ behind schedule

Cost variance = BCWP – ACWP
= \$70,000 – \$95,000
= –\$25,000 $\therefore$ over budget

THE CORRECT ANSWER IS: (B)

512. Reference: Kosmatka, Kerkhoff, and Panarese, *Design and Control of Concrete Mixtures,* 14th ed., Chapter 9.

Work on the basis of 1 yd^3 of concrete. The concrete consists of cement, fine aggregate, coarse aggregate, and water.

$$\text{Weight of cement} = (6.28 \text{ sacks/yd}^3)(94 \text{ lb/sack}) = 590.3 \text{ lb/yd}^3$$

$$\text{Volume of cement} = (590.3 \text{ lb/yd}^3)(1/3.15)(1 \text{ ft}^3/62.4 \text{ lb}) = 3.00 \text{ ft}^3/\text{yd}^3$$

$$\text{Vol of fine aggregate} = (2.25)(590.3 \text{ lb/yd}^3)(1/2.65)(1 \text{ ft}^3/62.4 \text{ lb}) = 8.03 \text{ ft}^3/\text{yd}^3$$

$$\text{Vol of coarse aggregate} = (3.25)(590.3 \text{ lb/yd}^3)(1/2.65)(1 \text{ ft}^3/62.4 \text{ lb}) = 11.60 \text{ ft}^3/\text{yd}^3$$

$$\text{Vol of water} = 27.0 - 3.0 - 8.03 - 11.60 = (4.37 \text{ ft}^3/\text{yd}^3)(7.48 \text{ gal/ft}^3) = 32.69 \text{ gal/yd}^3$$

$$\text{Water/cement ratio} = 32.69 \text{ gal/yd}^3/6.28 \text{ sacks/yd}^3 = 5.2 \text{ gal/sack}$$

THE CORRECT ANSWER IS: (D)

513. Reference: McCarthy, *Essentials of Soil Mechanics and Foundations,* 6th ed., Chapter 16.

Use at-rest pressure,

$$F_1 = (1/2)(45)(4)^2 = 360 \text{ lb/ft}$$

$$F_2 = (45)(4)(9) = 1,620 \text{ lb/ft}$$

$$F_3 = (1/2)(35 + 62.4)(9)^2 = 3,945 \text{ lb/ft}$$

$$\Sigma F = F_1 + F_2 + F_3 = 360 + 1,620 + 3,945 = 5,925 \text{ lb/ft}$$

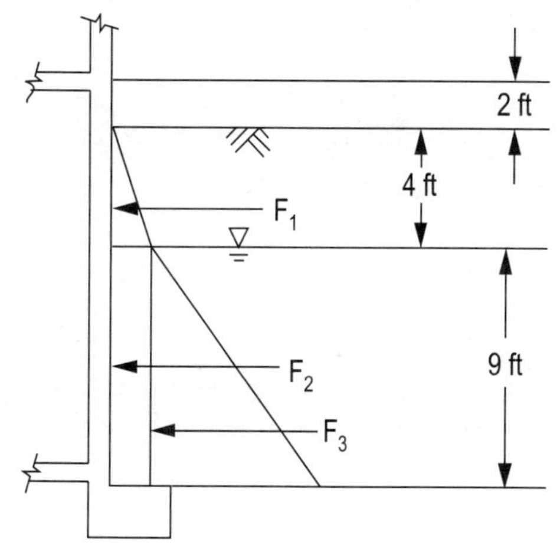

THE CORRECT ANSWER IS: (D)

514. Reference: ACI 347-04, Section 2.2.4.

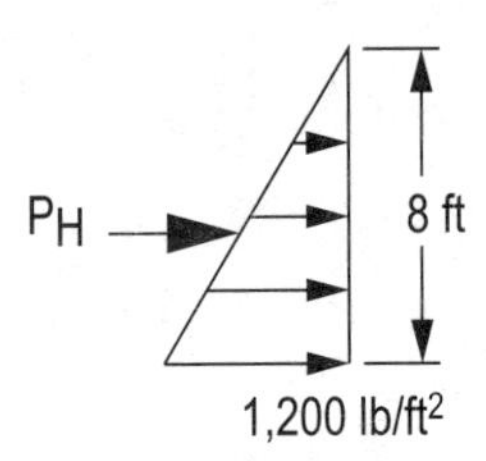

PRESSURE DIAGRAM
ON VERTICAL FACE

$$P_H = (1/2)(8\text{ ft})(1{,}200\text{ lb/ft}^2) = 4{,}800\text{ lb/ft}$$

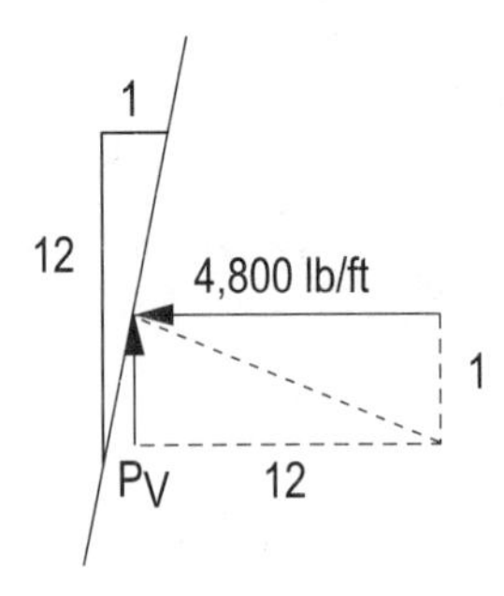

$$P_V = \frac{4{,}800\text{ lb/ft}}{12} = 400\text{ lb/ft}$$

THE CORRECT ANSWER IS: (C)

515. Reference: Meriam and Kraige, *Engineering Mechanics,* Vol. 1, 3rd ed., Chapter 3.

$$\Sigma M_A = 0$$

$$(12\text{ kips})(3\text{ in.}) - T(8\text{ in.}) = 0$$

$$T = \frac{(12\text{ kips})(3\text{ in.})}{8\text{ in.}}$$

$$= 4.5\text{ kips}$$

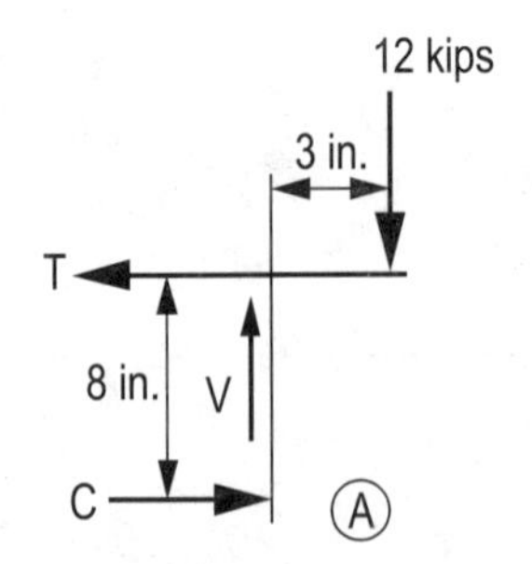

FREE-BODY DIAGRAM

THE CORRECT ANSWER IS: (B)

516. OSHA 1926.451 (c)(1) reads in part:

> Supported scaffolds with a height to base width (including outrigger supports, if used) ratio of more than four to one (4:1) shall be restrained from tipping by guying, tying, bracing, or equivalent means....

THE CORRECT ANSWER IS: (C)

517. Reference: Peurifoy, Schexnayder, and Shapira, *Construction Planning, Equipment, and Methods*, 7th ed., pp. 104–105.

Option (A) is correct: it is **NOT** true. On nuclear gages requiring a badge, the operator must also be certified. A safety course must be completed.

THE CORRECT ANSWER IS: (A)

518. Reference: OSHA Forms 300 and 300A.

The term *incidence rate* means the number of injuries and illnesses, or lost workdays, per 100 full-time workers. The rate is calculated as

$N \times 200{,}000/EH$

where

N = number of injuries and illnesses, or number of lost workdays.

EH = total hours worked by all employees during a month, a quarter, or a fiscal year.

200,000 = base for 100 full-time equivalent workers (including 40 hours per week, 50 weeks per year).

In this case $N = 4 + 3 + 5 = 12$

The incidence rate = $12\ (200{,}000)/750{,}000 = 3.20$

THE CORRECT ANSWER IS: (C)

519. Reference: Nunnally, *Construction Methods and Management,* 3rd ed., pp. 22–27.

The sample has the following properties:

% passing the #4 sieve	77%
% fines passing the #200 sieve	18%
% retained on the #200 sieve	100% – 18% = 82%
Liquid Limit, LL	32%
Plastic Limit, PL	25%
Plasticity Index, PI = LL – PL	7%

Based on 77% passing the #4 sieve and 82% retained on the #200 sieve, the soil is classified as a sand, either SM or SC. Based on the fines having LL = 32 and PI = 7, the fines would be classified as ML, nonplastic. Therefore, according to the Unified Soil Classification System, the sample is classified as SM.

THE CORRECT ANSWER IS: (C)

520. References: Cheney and Chassie, *Soils and Foundations Workshop Manual;* and Frank R. Walker Company, *Walker's Building Estimator's Reference Book*, 28th ed., pp. 240–259.

The existing ground elevation from the boring log is elevation 990. This puts the rock elevation at 929.5 ft. If the bottom of the pile cap is at elevation 980, the piles need to be 50.5 ft long, plus embedment in the pile cap, plus an allowance for damage during driving.

50.5 + 1.0 + 2.0 = 53.5 ft

THE CORRECT ANSWER IS: (B)

GEOTECHNICAL
AFTERNOON SOLUTIONS

Correct Answers to the GEOTECHNICAL Afternoon Sample Questions

Detailed solutions for each question begin on the next page.

501	A
502	D
503	B
504	C
505	C
506	D
507	C
508	D
509	B
510	B
511	A
512	D
513	C
514	D
515	C
516	A
517	C
518	B
519	C
520	B

501. $G = 2.65$

$e = 0.72$

$\gamma_w = 62.4 \text{ lb/ft}^3$

$$\gamma_b = \frac{(G-1)}{1+e}\gamma_w = \frac{(2.65-1)}{1.72}\left(62.4 \text{ lb/ft}^3\right) = 60 \text{ lb/ft}^3$$

THE CORRECT ANSWER IS: (A)

502. A dilatometer and a pressuremeter do not obtain samples. An SPT obtains a sample too distorted for testing.

THE CORRECT ANSWER IS: (D)

503. Compute the volume of the sample:

$$V = \frac{\pi d^2 h}{4} = \frac{\pi \times (3\text{ in.})^2 \times (6\text{ in.})}{4} = 42.4\text{ in}^3 = 0.0245\text{ ft}^3$$

Compute the volume of solids:

$$V_s = \frac{W_s}{G_s \gamma_w} = \frac{2.54\text{ lb}}{2.65 \times 62.4\text{ pcf}} = 0.0154\text{ ft}^3$$

Compute the volume of voids:

$$V_v = V - V_s = 0.0245 - 0.0154 = 0.0091\text{ ft}^3$$

Compute the void ratio:

$$e = \frac{V_v}{V_s} = \frac{0.0091}{0.0154} = 0.59$$

Compute the moisture content:

$$w = \frac{W_w}{W_s} = \frac{2.95\text{ lb} - 2.54\text{ lb}}{2.54\text{ lb}} = 0.161\text{ or }16.1\%$$

Compute the degree of saturation:

$$S = \frac{G_s w}{e} = \frac{2.65 \times 0.161}{0.59} = 0.72\text{ or }72\%$$

THE CORRECT ANSWER IS: (B)

504. $\sigma_1 = 33.5 \text{ psi}, \quad \sigma_3 = 16.4 \text{ psi}, \quad u = 10.0 \text{ psi}$

$$\sigma_1' = \sigma_1 - u = 23.5 \text{ psi}, \quad \sigma_3' = \sigma_3 - u = 6.4 \text{ psi}$$

$$p = \frac{\sigma_1' + \sigma_3'}{2} = \frac{23.5 + 6.4}{2} = 14.95 \text{ psi}$$

$$q = \frac{\sigma_1' - \sigma_3'}{2} = \frac{23.5 - 6.4}{2} = 8.55 \text{ psi}$$

$$\alpha' = \arctan\left(\frac{q}{p}\right) = \arctan\left(\frac{8.55}{14.95}\right) = 29.8°$$

$$\phi' = \arcsin(\tan\alpha) = \arcsin(\tan 29.8°) = 34.9°$$

THE CORRECT ANSWER IS: (C)

505. The seepage loss under the dam may be computed using the flow net as follows:

$$Q_1 = kHL\left(\frac{N_f}{N_d}\right) \text{ where:}$$

N_f = number of flow channels

N_d = number of equipotential drops

$$Q_1 = (0.003 \text{ fps})(12 \text{ ft})(120 \text{ ft})\left(\frac{6}{15}\right)$$

$$= 1.73 \text{ cfs}$$

THE CORRECT ANSWER IS: (C)

506. $V_v = V_t - V_s = 15.5 - 9.26 = 6.24 \text{ cm}^3$

$$V_s = \frac{W_s}{G\gamma_w} = \frac{25}{2.7 \times 1} = 9.26 \text{ cm}^3$$

$$e = \frac{V_v}{V_s} = \frac{6.24}{9.26} = 0.67$$

$$w = \frac{W_w}{W_s} = \frac{30.5 - 25}{25} \times 100 = 22\%$$

$Se = Gw$

$$S = \frac{2.7 \times 0.22}{0.67} = 0.89$$

THE CORRECT ANSWER IS: (D)

507. Determine $\Delta\sigma$ from foundation in chart shown in Question 507.

From chart, $x = 15$ $y = 15$ $z = 15$

so $m = 1$ $n = 1$

$I = 0.18$ Total $I_T = 4(I) = 4(0.18) = 0.72$

$P = 500$ $\Delta\sigma = 0.72(500) = 360$ psf

$\sigma'_{vo} = 5(110) + 5(110 - 62.4) + 5(95 - 62.4) = 951$ psf

$$S = C_c \frac{H_o}{1 + \ell_o} \log\left(\frac{\sigma'_{vo} + \Delta\sigma}{\sigma'_{vo}}\right)$$

$$S = 0.29 \frac{10(12)}{1 + 1.0} \log\left(\frac{951 + 360}{951}\right)$$

$S = 2.4$

THE CORRECT ANSWER IS: (C)

508. Well-graded sand is the least susceptible to frost heave.

THE CORRECT ANSWER IS: (D)

509. Pile end bearing capacity by Terzaghi equation.

$Q_{PB} = \sigma'_v N_q$ Arca

$\sigma'_v = 3(100) + 4(100 - 62.4) + 6(120 - 62.4) = 796 \text{ psf}$

$N_q = 260$ (given) $\text{Area} = \frac{\pi D^2}{4} = \frac{\pi(1)}{4} = 0.79 \text{ ft}^2$

$Q_{PB} = 796(260)(0.79) = 163{,}000 \text{ lb}$

$Q_{PB} = 82 \text{ tons}$

THE CORRECT ANSWER IS: (B)

510. If no movement is allowed, the retaining system should be designed using K_O.

THE CORRECT ANSWER IS: (B)

511. $\sigma'_{vo} = 5(125) + 2(110) + 2(110 - 62.5) + 5(120 - 62.5) = 1{,}228 \text{ psf}$

THE CORRECT ANSWER IS: (A)

512. Determine the pullout force in the bottom panel.

$$P = 0.30\left(200 \text{ psf} + 115 \text{ pcf} \times \frac{9 \text{ ft} + 12 \text{ ft}}{2}\right) 3 \text{ ft} \times 3 \text{ ft}$$

$$P = 3{,}800 \text{ lb}$$

$$P \times FS = L_e \times b \times \sigma_v \times \tan\delta \times \text{number of sides}$$

$$3{,}800 \times 3 = L_e \times 1 \times 115\left(\frac{9+12}{2}\right) \times \tan 22^\circ \times 2$$

$$L_e = 11.8 \text{ ft} \qquad \text{Use 12 ft.}$$

THE CORRECT ANSWER IS: (D)

513. Soil Layer 1

$$Gw = Se \qquad 2.68 \times 0.123 = S \times 0.75$$

$$S = 0.439$$

$$\gamma_{wet} = \frac{\gamma_w G(1+w)}{1+e} = \frac{62.4 \times 2.68(1+0.123)}{1+0.75} = 107.3 \text{ pcf}$$

Soil Layer 3

$$Gw = \not{S}^{1.0} e \qquad e = 2.70 \times 0.25 = 0.675$$

$$\gamma_{sat} = \frac{\gamma_w G(1+w)}{1+e} = \frac{62.4 \times 2.70\,(1+0.25)}{(1+0.675)} = 125.7 \text{ pcf}$$

Effective vertical stress at middle of Layer 3

$$\sigma'_v = (107.3 \times 10) + (112 \times 5) + (112 - 62.4) \times 3 + (125.7 - 62.4) \times 3.5 = 2{,}003 \text{ psf}$$

Shear stress to cause liquefaction

$$\Sigma' = \sigma'_v \times CSR = 2{,}003 \times 0.29 = 580.9 \text{ psf}$$

Determine factor of safety against liquefaction in Layer 3

$$FS = \frac{\Sigma'}{\Sigma} = \frac{580.9}{450} = 1.29$$

THE CORRECT ANSWER IS: (C)

514. (A) is incorrect because the "flowing artesian" effect applies to wells not aquifers, and in any event in the situation described, the piezometric surface of the deep aquifer is below the ground surface and not flowing out above the ground surface.

(B) is incorrect because the aquifer is confined by the upper clay layers within a piezometric level well above the bottom of the clay layer.

(C) obviously does not apply to a coarse sand and gravel aquifer.

THE CORRECT ANSWER IS: (D)

515. $\eta H = 20$

$H = 20$

$\eta = 1.0$

From the Taylor Chart,

$$Sn \simeq 0.169 = \frac{c_r}{\gamma H}$$

$$0.169 = \frac{c_r}{(120)(20)}$$

$$c_r = 405.6$$

$$FS = \frac{c}{c_r} = \frac{750}{405.6} = 1.85$$

THE CORRECT ANSWER IS: (C)

516. $Q_{ult} = c\,N_c\,S_c\,d_c + \gamma\,D_f\,N_q\,S_q + \frac{1}{2}\gamma\,B\,N_\gamma S_\gamma$

$$= 200(14.8)\left(1+\frac{N_q}{N_c}\right)\left(1+\frac{0.2(2)}{5}\right) + 115(2)(6.4)(1+\tan 20°) + \frac{1}{2}(115)(5)(5.4)(0.6) = 7{,}503 \text{ psf}$$

$$Q_{all} = \frac{Q_{ult}}{FS} = \frac{7{,}503}{3} = 2{,}500 \text{ psf}$$

THE CORRECT ANSWER IS: (A)

517. Overturning $FS = \dfrac{\text{Resisting moment}}{\text{Driving moment}}$

Resisting moment = (2)(9)(155)(4.5) + (5)(18)(155)(4.5) + (2)(18)(105)(1) = 79,110 ft-lb

Active force $P_a = \dfrac{1}{2} K_a \, \gamma \, H^2$

$$K_a = \tan^2\left(45 + \frac{\phi}{2}\right) = 0.33$$

$$P_a = \frac{1}{2} \times 0.33 \times 105 \times 20^2 = 6{,}930 \text{ lb}$$

$$\text{Driving moment} = P_a \times \frac{H}{3} = 6{,}930 \times \frac{20}{3} = 46{,}200 \text{ ft-lb}$$

$$FS = \frac{79{,}110}{46{,}200} = 1.71$$

THE CORRECT ANSWER IS: (C)

518. $T_{all} = \dfrac{T_{ult}}{FS} + W_{pile}$

$T_{ult} = K_H \, P_o \, \tan\delta \times H \times S$

Surface area, $S = 2 \text{ ft} \times 4 = 8$ ft/lf

Pile embedment, $H = 30$ ft

$\tan\delta = \tan(0.75 \times 32°) = 0.445$

$$\text{Average overburden pressure, } P_o = \frac{5(120) + 5(120) + 30(115)}{2} = 2{,}325 \text{ psf}$$

$T_{ult} = 1.0 \times 2{,}325 \times 0.445 \times 30 \text{ ft} \times 8 \text{ ft} = 248$ kips

$$W_{pile} = 2 \text{ ft} \times 2 \text{ ft} \times \frac{150}{1{,}000} \times 35 = 21 \text{ kips}$$

$$T_{all} = \frac{248}{3} + 21 = 103.7 \text{ kips}$$

THE CORRECT ANSWER IS: (B)

519. Option (A): An increase in the relative density of the soil in front of the wall will increase Kp, increasing the safety factor.

Option (B): Extending the wall will increase the safety factor.

Option (C): A sudden drop in water elevation will increase the likelihood of overturning, reducing the safety factor.

Option (D): A gradual drop in water elevation will result in a lower P_A, increasing the safety factor.

THE CORRECT ANSWER IS: (C)

520. The embankment requires 500,000 yd^3 of soil at:

$$\gamma_{dry} = (0.90)(120.0) = 108.0 \text{ pcf}$$

The total weight of dry soil required is:

$$W_{total} = (500,000 \text{ yd}^3)(27 \text{ ft}^3/\text{yd}^3)(108.0 \text{ pcf}) = 1.458 \times 10^9 \text{ lb}$$

The dry unit weight of soil in the truck is:

$$\gamma_{dry} = G_s \gamma_w/(1 + e) = (2.65)(62.4)/(1+1.30) = 71.9 \text{ psf}$$

Each truck can carry a weight of:

$$W_{truck} = (5.0 \text{ yd}^3)(27 \text{ ft}^3/\text{yd}^3)(71.9 \text{ pcf}) = 9,700 \text{ lb/truck}$$

Therefore, the minimum number of trucks required is:

$$W_{total}/W_{truck} = 1.458 \times 10^9 / 9,700 = 150,000 \text{ trucks}$$

THE CORRECT ANSWER IS: (B)

STRUCTURAL
AFTERNOON SOLUTIONS

Correct Answers to the STRUCTURAL Afternoon Sample Questions

Detailed solutions for each question begin on the next page.

501	A
502	C
503	D
504	C
505	B
506	D
507	C
508	B
509	C
510	A
511	C
512	D
513	A
514	C
515	A
516	A
517	C
518	D
519	D
520	D

501. **ASD**

Per p. 3-74, AISC *Steel Construction Manual*, 13th ed.

$V_n/\Omega_v = 62.5$ kips

Max shear

LRFD

Per p. 3-74, AISC *Steel Construction Manual*, 13th ed.

$\phi_v V_n = 93.7$ kips

THE CORRECT ANSWER IS: (A)

502. $p_f = 0.7\ C_e\ C_t\ I\ p_g$ (Eq. 7-1)

where $C_e = 0.9$ (Table 7-2)

$C_t = 1.2$ (Table 7-3)

$I = 0.8$ (Table 7-4)

$p_g = 20$ psf

$p_f = (0.7)(0.9)(1.2)(0.8)(20)$

$= 12.1$ psf

$p_s = C_s p_f$ (Eq. 7-2) and $C_s = 1.0$ (Fig 7-2)

$= 1.0(12.1) = 12.1$

Unbalanced snow load $= I\ p_g$ (Paragraph 7.6.1)

$= 0.8(20)$

$= 16$ psf

16 psf controls

THE CORRECT ANSWER IS: (C)

503. M_{OT} = 30 kips × 100 ft = 3,000 ft-kips

T = 3,000 ft-kips/30 ft = 100 kips/2 sides = 50 kips per side

Maximum uplift occurs when tank is empty.

Tank = 70 kips

Tower frame = 0.4 klf × 90 ft = 36 kips

Total weight = 70 kips + 36 kips = 106 kips

Weight per leg = 106 kips/4 = 26.5 kips

Net uplift = –50 kips + 26.5 kips

= 23.5 kips

THE CORRECT ANSWER IS: (D)

504. $M_u = \phi M_n$

$\phi = 0.90$ per ACI 318

$M_n = 0.85\, f'_c\, b_w a(d - a/2)$

$$\frac{(80 \text{ ft-kips})(12 \text{ in./ft})}{0.90} = 0.85(4 \text{ ksi})(12 \text{ in.})(a)(27.5 \text{ in.} - a/2)$$

$1,067 = 40.8(a)(27.5 - a/2)$

$1,067 = 1,122a - 20.4a^2$

$a = 0.97$

$0.85\, f'_c\, b_w a = A_s f_y$

$0.85(4 \text{ ksi})(12 \text{ in.})(0.97 \text{ in.}) = A_s(60 \text{ ksi})$

$A_s = 0.66 \text{ in}^2$

Check minimum area of steel per ACI 318 10.5.1:

$$A_{s_{min}} = \frac{200\, b_w d}{f_y} = \frac{200(12)(27.5)}{60,000} = 1.10 \text{ in}^2$$

ACI 10.5.3 allows $A_s < A_{s_{min}}$ if A_s provided $> \frac{4}{3} A_s$ required.

$$\frac{4}{3}(0.66 \text{ in}^2) = 0.88 \text{ in}^2$$

THE CORRECT ANSWER IS: (C)

505. Reference: ACI 530, 2.3.5.3.

$$A_v = \frac{VS}{F_s d}$$

$$V = \frac{A_v F_s d}{S} = \frac{0.2 \times 24,000 \times 20}{8} = 12.0 \text{kips}$$

THE CORRECT ANSWER IS: (B)

506. **ASD provisions:**
Part 5, p. 5-1
Table 5-8
pp. 5-48, 5-49
D2, p. 16.1-26

Calculate the allowable tensile stresses on the gross area.

$$R_o = P_n/\Omega t$$

$$F_t = 0.6\ F_y = 0.6 \times 36\ \text{ksi} = 21.6\ \text{ksi}$$

Calculate the minimum gross area.

$$A_g = P/F_t = 85\ \text{kips}/21.6\ \text{ksi} = 3.94\ \text{in}^2$$

Select highest-weight double-angle size.
From p. 1-42
2L 4 × 4 × 3/8 (2 × 9.8 = 19.6 lb/ft) has a gross area of (2 × 2.86 = 5.72 in^2)
5.72 in^2 > 3.94 in^2 OK, but check net area

Calculate the allowable tensile stresses on the effective net area. (D2-2, p. 16.1-27)

$$F_t = 0.5\ F_u = 0.5 \times 58\ \text{ksi} = 29.0\ \text{ksi}$$

Calculate the net area. (Table 9-1, p. 9-21)

$$A_n = A_g - A_{holes} = 5.72\ \text{in}^2 - 2 \times (3/8\ \text{in.})(7/8\ \text{in.} + 1/16\ \text{in.}) = 5.02\ \text{in}^2$$

Calculate the effective net area. (D3.3)
Three bolts. Therefore U = 0.60 (Table D3.1, pp. 16.1-29)

$$A_e = U\ A_n = 0.60 \times 5.02\ \text{in}^2 = 3.01\ \text{in}^2$$

Calculate the actual tensile stress on the effective net section.

$$f_t = P/A_e = 85\ \text{kips}/3.01\ \text{in}^2 = 28.3\ \text{ksi} < 29.0\ \text{ksi} \qquad \text{OK}$$

Select 2L 4 × 4 × 3/8

LRFD provisions:
Chapter D, p. 16.1-26
Section D2

Calculate the design tensile strength on the gross area.

$$\phi_t P_n = \phi_t\ F_y A_g = 0.9 \times 36\ \text{ksi} \times A_g \geq 130\ \text{kips} \qquad \text{(D2-1)}$$

Calculate the minimum gross area.

$$A_g = 130\ \text{kips}/(0.9 \times 36\ \text{ksi}) = 4.01\ \text{in}^2$$

506. (Continued)

Select least-weight double-angle size.

From AISC, 13th ed., p. 1-104

2L 4 × 4 × 3/8 (19.6 lb/ft) has a gross area of 5.72 in^2 > 4.01 in^2 OK, but check net area.

Calculate the design tensile strength on the effective net area. (Table 9-1, p. 9-21)

See effective net area calculations under the provisions and Chapter D2, D3.

$$\phi_t P_n = \phi_t F_u A_e = 0.75 \times 58 \text{ ksi} \times 3.01 \text{ in}^2 = 130.9 \text{ kips} > 130 \text{ kips} \quad \text{OK}$$

THE CORRECT ANSWER IS: (D)

507. $\Delta_{vA} = \Delta_{vB}$

$$\Delta_{vA} = V_A (40 \text{ ft}) / [(60 \text{ ft})(30 \text{ kips/ft})]$$

$$= 0.022\ V_A$$

$$V_A + V_B = 100$$

$$V_B = 100 - V_A$$

$$\Delta_{vB} = [(100 \text{ kips} - V_A)(70 \text{ ft})] / [(100 \text{ ft})(50 \text{ kips/ft})]$$

$$= 1.4 - 0.014\ V_A$$

$$0.022\ V_A = 1.4 - 0.014\ V_A$$

$$0.036\ V_A = 1.4$$

$$V_A = 1.4/0.036$$

$$= 38.9 \text{ kips}$$

THE CORRECT ANSWER IS: (C)

508. **ASD solution:**

Allowable tension load

$R_n = F_w A_w$ (J2-3)

$F_w = 0.60\ Fe_{xx}$ (Table J2.5)

$\Omega = 2.0$ (Table J2.5)

$$\frac{R_n}{\Omega} = \frac{(0.60\ Fe_{xx})(0.707\ S_w)L_w}{2.0}$$

$$= (0.60 \times 70)(0.707 \times 0.25)[2(9 + 0.75)]$$

$$= 72.38 \text{ kips}$$

$R = 35$ kips, shear load given

Find resultant load for the two forces

$$T_a = \sqrt{(R_n/\Omega)^2 - R^2}$$

$$T_a = \sqrt{(72.38)^2 - (35)^2}$$

$$T_a = 63.4 \text{ kips}$$

LRFD solution:

Allowable tension load

$R_n = F_w A_w$ (J2-3)

$F_w = 0.60\ Fe_{xx}$ (Table J2.5)

$\phi = 0.75$ (Table J2.5)

$$\phi R_n = 0.75(0.60\ Fe_{xx})(0.707\ S_w)L_w$$

$$= 0.75(0.60 \times 70)(0.707 \times 1/4)[(2)(9 + 0.75)]$$

$$= 108.6 \text{ kips}$$

$R = 45.5$ kips, shear load given

Find resultant load for the two forces

$$T_a = \sqrt{(\phi R_n)^2 - R^2}$$

$$T_a = \sqrt{(108.6)^2 - (45.5)^2}$$

$$T_a = 98.6 \text{ kips}$$

THE CORRECT ANSWER IS: (B)

509. Weight = (410 in^2/144 in^2/ft^2) (0.110 kcf) = 0.31 klf

M = (0.31 klf) (60 ft)^2/8 = 139.5 ft-kips (positive moment)

At end: F_{ps} = (8) (0.153 in^2) (195 ksi) = 239 kips

M_{ps} = (239 kips) (9.5 in.)/(12 in./ft) = 189 ft-kips (negative moment)

At center: M_{ps} = (239 kips) (14.25 in.)/(12 in./ft) = 284 ft-kips (negative moment)

M_{tot} = 284 – 139.5 = 144.50 ft-kips

Moment is negative at ends, negative at center

THE CORRECT ANSWER IS: (C)

510. P_{tot} = 600 kips

Bearing pressure = 600 kips/(10 ft × 22 ft) = 2.73 ksf

d = 36 in. (given)

Critical bending shear, V = 2.73 ksf (8 ft – 1 ft – 3 ft)(10 ft)

= 109.2 kips

THE CORRECT ANSWER IS: (A)

511. $P_{roof} = 12(30) = 360 \text{ plf}$

$e = \frac{7.625}{2} + 3.5 = 7.31 \text{ in.}$

$M_{roof} = 360(7.31)/12 = 219 \text{ ft-lb/ft}$

M_{wind} @ midheight $= 20(12)^2/8 = 360$ ft-lb/ft

$\Sigma M_{midheight} = 360 + 219/2 = 480$ ft-lb/ft

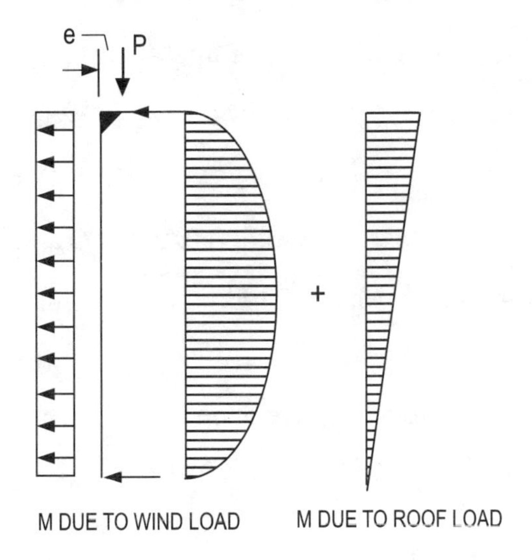

Alternative Solution:

Alternative couple (shear): $\frac{219}{12} = 18.2, \quad 20 \times \frac{12}{2} = 120$

Superimpose shears (top): 18.2 + 120 = 138.2

$V = 0 \text{ @ } \frac{138.2}{20} = 6.9$ ft from top

$\therefore$ M @ 6.9 ft $= 138.2 \times \frac{6.9}{2} = 477$ ft-lb

THE CORRECT ANSWER IS: (C)

512. $\varnothing = \frac{TL}{JG}$

For the 2-in.-dia. shaft:

$T = 5 \text{ ft-kips}, \quad L = 36 \text{ in.}, \quad J = \frac{\pi d^4}{32} = 1.6 \text{ in}^4$

$$\varnothing = \frac{5 \text{ ft-kips}(12 \text{ in.})(36 \text{ in.})}{(4,400 \text{ kips/in}^2)(1.6 \text{ in}^4)} = 0.31 \text{ rad}$$

For the 3-in.-dia. shaft:

$T = 15 + 5 = 20 \text{ ft-kips}, \quad L = 60 \text{ in.}, \quad J = 8 \text{ in}^4$

$$\varnothing = \frac{20 \text{ ft-kips}(12 \text{ in.})(60 \text{ in.})}{(4,400 \text{ kips/in}^2)(8 \text{ in}^4)} = 0.41 \text{ rad}$$

Total angle of twist $= 0.31 + 0.41 = 0.72$ rad

THE CORRECT ANSWER IS: (D)

513. Area of flange = 6 in. × 0.5 in. = 3 in^2

$Xi_{flange} = 5.75$ in.

$Q_{flange} = 3 \text{ in}^2 (5.75 \text{ in.}) = 17.25 \text{ in}^3$

$$v = \frac{VQ}{I} = \frac{20 \text{ kips}(17.25 \text{ in}^3)}{232 \text{ in}^4} = 1.5 \text{ kips/in.}$$

THE CORRECT ANSWER IS: (A)

514. $f_v = \dfrac{3V}{2bd}$ for a rectangular cross section

$b = 2$ in.
$d = 12$ in.

V = maximum shear force—all loads within a distance from supports equal to the depth, d, of the bending member are ignored.

$$f_v = \frac{(3)\left(\dfrac{18\text{ ft}}{2} - 1\text{ ft}\right)(60\text{ lb/ft})}{(2)(2\text{ in.})(12\text{ in.})} = 30.0\text{ psi}$$

THE CORRECT ANSWER IS: (C)

515. Reference: NDS 2005.

Eq. 3.10-1 $\quad F'_\theta = \dfrac{F_c^* F'_{c\perp}}{F_c^* \sin^2\theta + F'_{c\perp}\cos^2\theta}, \quad \theta = \tan^{-1}\left(\dfrac{12}{5}\right) = 67.38°$

$$F_c^* = C_D C_M C_t C_F C_i F_c$$

$$F_{c\perp} = C_M C_t C_i C_b F_{c\perp}$$

Table 4A NDS Supplement: $F_c = 1{,}350$ psi, $F_{c\perp} = 405$ psi

Part 4: $C_D = C_M = C_t = C_F = C_b = C_i = 1.0$

(C_D given, Dry, Normal Temp., Table 4A, End of Member, Not Incised)

$$F_c^* = (1.0)(1.0)(1.0)(1.0)(1.0)(1{,}350) = 1{,}350\text{ psi}$$

$$F_{c\perp} = (1.0)(1.0)(1.0)(1.0)(405) = 405\text{ psi}$$

$$F'_\theta = \frac{(1{,}350)(405)}{1{,}350[\sin^2(67.38)] + 405[\cos^2(67.38)]} = \frac{546{,}750}{1{,}150 + 60} = 452\text{ psi}$$

THE CORRECT ANSWER IS: (A)

516. Reference: PCI, Chapter 4.

Total prestress force = 2 × 6 strands × 0.153 in^2/strand × 175 ksi = 321.3 kips

Axial stress = P/A = 321.3/(36 × 18) = 0.496 ksi (compression)

Bending stress due to prestress = Pe/S $\quad$ $S = bd^2/6 = 18 \times 36^2/6 = 3{,}888\ in^3$
$e = 36/2 - 3 - 2/2 = 14$ in.

Bending stress due to prestress = 321.3 × 14/3,888 = 1.158 ksi (tension at top)

Bending stress due to beam self-weight = 0.65 ksi (compression at top)

Thus, stress at top at release = 0.49 – 1.15 + 0.65 = –0.01 ksi (tension)

THE CORRECT ANSWER IS: (A)

517. Reference: AASHTO Section 9.7.2.3.

$S = 8.5\ ft - b_f + overhang$

$= 8.5\ ft - 1.5\ ft + 0.5\ ft$

$= 7.5\ ft$

THE CORRECT ANSWER IS: (C)

518. Reference: AISC (ASD or LRFD), 13th ed.

See p. 16.1-160, Section 3.3, Design Stress Range

Eq. A-3-1 $$F_{SR} = \left(\frac{C_f}{N}\right)^{0.333} \geq F_{TH}$$

Table A-3.1 gives (for rolled shapes, Stress Category A)

$C_f = 250\ (10^8)$

$F_{TH} = 24$ ksi

$$F_{SR}\left(\frac{250(10^8)}{48[3(5)(25)]}\right)^{0.333} = (57,077)^{0.333} = 37.1\text{ ksi} > F_{TH} = 24\text{ ksi}$$

THE CORRECT ANSWER IS: (D)

519. Overturning moment, $M_{ot} = 20\text{ kips} \times 22\text{ ft} = 440\text{ ft-kips}$

Restoring moment, $M_r = 80\text{ kips} \times 2\text{ ft} + \dfrac{(2.5\text{ kips/ft } \times 24\text{ ft})^2}{2} = 880\text{ ft-kips}$

Footing weight, $W_{ftg} = 2.5\text{ kips/ft} \times 24\text{ ft} = 60\text{ kips}$

Total vertical load, $P_{tot} = 80\text{ kips} - 20\text{ kips} + 60\text{ kips} = 120\text{ kips}$

Resultant location, $y = \dfrac{M_r - M_{ot}}{P_{tot}} = \dfrac{880\text{ ft-kips} - 440\text{ ft-kips}}{120\text{ kips}} = 3.7\text{ ft}$

Since resultant is outside the middle third of the footing base:

$$P_{max} = \frac{2}{3} \times \frac{(P_{tot})}{(y \times \text{footing width})} = \frac{2\ (120\text{ kips})}{3\ (3.7\text{ ft} \times 8\text{ ft})} = 2.7\text{ ksf}$$

THE CORRECT ANSWER IS: (D)

520. The ultimate pile capacity, Q_u, is calculated by the following equation:

$Q_u = Q_{design}$ (F.S.)

$Q_u = 50 \text{ kips} \times 2.5 = 125 \text{ kips}$

$Q_u = Q_f + Q_s = 125 \text{ kips}$

Skin Friction

$Q_f = \alpha(C)(P)(L)$

$C = q_u/2 = 1{,}100 \text{ psf}$

$\alpha = 0.78$

$P = 2\pi r = (2)(\pi)(7.5/12) = 3.92 \text{ ft}$

End Bearing

$Q_s = 0$

Find Length of Pile Required

125 kips (1,000 lb/kip) = 0.78 (1,100 psf)(3.92 ft)(L)

L = 37 ft

THE CORRECT ANSWER IS: (D)

TRANSPORTATION
AFTERNOON SOLUTIONS

Correct Answers to the TRANSPORTATION Afternoon Sample Questions

Detailed solutions for each question begin on the next page.

501	A
502	C
503	D
504	B
505	B
506	A
507	C
508	D
509	B
510	C
511	B
512	C
513	C
514	B
515	A
516	D
517	D
518	A
519	C
520	B

501. Reference: *Highway Capacity Manual*, 2000, pp. 16-5, 16-48, and 16-49.

$$G_p = 3.2 + \left(\frac{L}{S_p}\right) + (0.27N_{ped}) \qquad \text{(for } W_E \leq 10 \text{ ft)}$$

$$= 3.2 + \frac{72}{4} + [(0.27)(0.9)]$$

$$= 21.4 \text{ sec}$$

THE CORRECT ANSWER IS: (A)

502. Reference: *Highway Capacity Manual*, 2000, Table 23-2.

The maximum service flow rate per lane for Level of Service D and a free-flow speed of 70 mph is 2,150 pcphpl.

THE CORRECT ANSWER IS: (C)

503. Let X = total number of vehicles.

(0.85)(X)(2) = total number of axles on passenger cars (2 axles)

(0.10)(X)(3) = total number of axles on 3-axle trucks

(0.03)(X)(4) = total number of axles on 4-axle trucks

(0.02)(X)(5) = total number of axles on 5-axle trucks

24,560,000 axles = (0.85)(X)(2) + (0.10)(X)(3) + (0.03)(X)(4) + (0.02)(X)(5)

24,560,000 axles = 2.22 X

X = 11,063,063 vehicles

Compute average annual daily traffic (AADT)

AADT = total number of vehicles per year/days per year

= 11,063,063/365 = 30,310 vehicles

THE CORRECT ANSWER IS: (D)

504. Reference: AASHTO, *A Policy on Geometric Design of Highways and Streets,* 2004, p. 273.

Determine the stopping sight distance.

For $S < L$

$$L = \frac{AS^2}{400 + 3.5S}$$

where S = stopping sight distance

A = algebraic difference in grades

L = length of curve

$$900 = \frac{8S^2}{400 + 3.5S}$$

Solve for S

$S = 486.285$ ft

THE CORRECT ANSWER IS: (B)

505. Calculate the present worth for the annual maintenance.

$$P_{AM} = \text{(annual maintenance for first 10 yr) (P/A, 10\%, 10 yr)}$$
$$+ \text{(annual maintenance for second 10 yr) (P/A, 10\%, 10 yr) (P/F, 10\%, 10 yr)}$$
$$= \$50{,}000\ (6.1446) + \$75{,}000\ (6.1446)\ (0.3855) = \$307{,}000 + \$178{,}000 = \$485{,}000$$

Calculate the present worth for the major maintenance.

$$P_{MM} = \text{(major maintenance) (P/F, 10\%, 10 yr)}$$
$$= \$300{,}000\ (0.3855) = \$116{,}000$$

Calculate the present worth of the residual value.

$$P_{RV} = \text{(residual value) (P/F, 10\%, 20 yr)}$$
$$= \$3{,}000{,}000\ (0.1486) = \$446{,}000$$

Calculate the sum, noting that the present worth of the first cost P_{FC} = \$6,000,000 and that the residual value P_{RV} is a deduction.

$$\text{Present worth} = P_{FC} + P_{AM} + P_{MM} - P_{RV}$$
$$= \$6{,}000{,}000 + 485{,}000 + 116{,}000 - 446{,}000$$
$$= \$6{,}155{,}000$$

THE CORRECT ANSWER IS: (B)

506. Determine the traffic volume in the year 2012.

Use the interest tables with an interest rate of 5% for 6 years; i.e., from 2006 to 2012.

$$V_{2012} = V_{2006}\ \text{(F/P, 5\%, 6 yr)} = 30{,}000\ \text{veh/day}\ (1.340) = 40{,}000\ \text{veh/day}$$

THE CORRECT ANSWER IS: (A)

507. Reference: ITE, *Traffic Engineering Handbook*, 5th ed., p. 332.

$L = VK25(1 + p)/Nc$

where:

L = Storage length (ft)
V = Peak 15-min flow rate (vph)
K = Constant to reflect random arrival of vehicles (usually 2)
Nc = Number of cycles per hour
p = Percentage of trucks or buses

The 25 is the length of a typical passenger car in feet.

$V = 35 \times 4 = 140$

$L = (140)(2)(25)(1 + 0.04)/60 = 121.33$ ft

THE CORRECT ANSWER IS: (C)

508. Reference: AASHTO, *Policy on Geometric Design of Highways and Streets*, 2004.

For a design speed of 50 mph, the sight distance S is 1,835 ft (p. 124, Exhibit 3-7 or p. 272, Exhibit 3-73)

The equation for the length of crest curve (AASHTO p. 270, Equation 3-45) when $S < L$ is:

$$L = AS^2/2{,}800$$

where $A = g_2 - g_1 = -2.5 - 1.0 = 3.5$

$$L = 3.5(1{,}835)^2 / 2{,}800 = 4{,}209.1 \text{ ft}$$

THE CORRECT ANSWER IS: (D)

509. Reference: AASHTO, *Roadside Design Guide,* 2002.

Using the curves and design data provided, the clear zone distance is 26 ft.

THE CORRECT ANSWER IS: (B)

510. Compute the rate of change of grade, r.

$$r = (g_2 - g_1) / L = [-3.0\% - (+4.5\%)] / 14 \text{ sta.} = -0.5357\% / \text{sta.}$$

Compute the distance from the PVC to the high point.

$$X_{PVC} = -g_1/r = -(+4.5\%)/(-0.5357\%/\text{sta}) = 8.4002 \text{ sta}$$

Compute the PVC station.

$$PVC = PVI - L/2 = (42+00) - (7+00) = 35+00$$

Compute the station of the high point.

$$\begin{aligned} \text{High point station} &= PVC + X_{PVC} \\ &= (35+00) + (8+40.02) \\ &= 43+40.02 \end{aligned}$$

THE CORRECT ANSWER IS: (C)

511. For the distance traveled on sandy soil before hitting the tree:

$$S_1 = V_1^2 - V_f^2 / 30f$$

$$25 = V_1^2 - (30)^2 /(30 \times 0.55)$$

$$V_1 = 36.2 \text{ mph}$$

Thus, the vehicle was traveling at 36.2 mph when it left the asphalt pavement.

Now repeat these calculations to determine the speed when braking began.

$$S_2 = V_2^2 - V_1^2 /(30 \times 0.71)$$

$$(150 + 24) = V_2^2 - (36.23)^2 /(30 \times 0.71)$$

$$V_2 = 70.8 \text{ mph}$$

THE CORRECT ANSWER IS: (B)

512. Reference: *Manual on Uniform Traffic Control Devices,* 2003, U.S. Department of Transportation—Federal Highway Administration, Washington, DC.

A shifting taper is at least 0.5 L. (Table 6C-3)

L = WS = 12(55) = 660 ft (Table 6C-4)

Shifting taper = 0.5 L = 330 ft minimum

THE CORRECT ANSWER IS: (C)

513. Reference: AASHTO, *Policy on Geometric Design of Highways and Streets*, 2004, Ex 9-103, p. 735.

$$d_H = AV_v t + \frac{(BV_v^2)}{a} + D + d_e$$

$$d_H = (1.47)(45)(2.5) + \left(\frac{1.075(45)^2}{11.2}\right) + 15 + 5$$

$$d_H = 380 \text{ ft}$$

THE CORRECT ANSWER IS: (C)

514. Reference: AASHTO, *Policy on Geometric Design of Highways and Streets*, 2004.

The rate of superelevation is given by the following equation:

$$0.01e + f = V^2/15R$$

where:

- e = Rate of roadway superelevation (%)
- f = Side friction factor
- V = Vehicle speed (mph)
- R = Radius of curve (ft)

$$0.01e + 0.12 = 60^2/(15)(1{,}091)$$

$$e = 10\%$$

THE CORRECT ANSWER IS: (B)

515. $Y = Y_{PVC} + g_1x + rx^2/2$

where:

Y	=	Curve elevation
Y_{PVC}	=	Elevation of PVC
g_1	=	Grade of the back tangent (%)
g_2	=	Grade of the forward tangent (%)
x	=	Horizontal distance from PVC to the point on the curve (in stations)
r	=	Constant that is the rate of change of grade $(g_2 - g_1)/L$
L	=	Length of vertical curve (in stations)

$$Y_{PVC} = (500)(0.2) + 103 = 113$$

$$x = 8.22$$

$$r = [3 - (-2)]/10 = 0.5$$

$$Y = 113 + (-2)(8.22) + [0.5(8.22)^2]/2 = 113.45 \text{ ft}$$

THE CORRECT ANSWER IS: (A)

516. Reference: AASHTO, *Policy on Geometric Design of Highways and Streets*, 2004.

The acceleration length given in Exhibit 10-70, p. 847, is 1,350 ft.

THE CORRECT ANSWER IS: (D)

517. The depth may be found using a trial and error method to solve Manning's equation (below) for the depth, y.

$$Q = \frac{1.486}{n} A R^{2/3} S^{1/2}$$

where:

A = area
R = hydraulic radius
S = slope = 0.5% = 0.005
n = 0.02

A = (b + zy)y where b is the bottom width and z = 3, since the side slopes are 3:1 (H:V).

$$R = \frac{(b+zy)y}{b+2y\sqrt{1+z^2}}$$

$$Q = \frac{1.486}{0.02} A R^{2/3} (0.005)^{1/2} = 30.0$$

Trial	Depth (in.)	Depth (ft)	R (ft)	A (ft^2)	Q (cfs)
1	12	1.00	0.601	5.00	18.7
2	15	1.25	0.726	7.188	30.5

The depth is most nearly 15 in.

THE CORRECT ANSWER IS: (D)

518. Reference: Brater and King, *Handbook of Hydraulics,* 6th ed., p. 6-15.

Solve Manning's equation for the diameter D.

$D = [2.159 \times Q \times n / (S)^{1/2}]^{3/8}$

$Q = 16$ cfs

$n = 0.012$

$S = 0.20\% = 0.002$

Then $D = [(2.159 \times 16 \times 0.012)/(0.002)^{1/2}]^{3/8}$

$= (9.268)^{3/8}$

$= 2.30 \text{ ft} \times 12 \text{ in./ft} = 27.65 \text{ in.}$

THE CORRECT ANSWER IS: (A)

519. For a 10-year storm and a time of concentration of 7.5 min:

I = (2.95 + 2.08)/2 = 2.5 in./hr

From the figure, C = 0.82

THE CORRECT ANSWER IS: (C)

520. Reference: *Standard Specifications for Transportation Materials and Method of Sampling and Testing,* 15th ed., pp. 152–154.

The percent passing the #200 sieve is greater than 35%. Therefore the soil is silt-clay.

LL = 56, PL = 47, therefore PI = 56 – 47 = 9

Meets A-5 specs (LL > 41, PI < 10)

Group index equation: $(F - 35)\,[0.2 + 0.005(LL - 40)] + 0.01(F - 15)\,(PI - 10)$

$$GI = (40 - 35)\,[0.2 + 0.005(56 - 40)] + 0.01(40 - 15)\,(9 - 10)$$

$$= 5(0.2 + 0.08) + 0.01(25)(-1) = 1.4 - 0.25 = 1.15$$

THE CORRECT ANSWER IS: (B)

WATER RESOURCES AND ENVIRONMENTAL AFTERNOON SOLUTIONS

Correct Answers to the WATER RESOURCES AND ENVIRONMENTAL Afternoon Sample Questions

Detailed solutions for each question begin on the next page.

501	B
502	B
503	A
504	B
505	D
506	C
507	D
508	B
509	B
510	D
511	B
512	D
513	C
514	B
515	D
516	B
517	B
518	A
519	C
520	A

501. Reference: Chen, *Civil Engineering Reference Handbook,* p. 1129.

The flow per connection is 1 gpm. Therefore, the total flow is 500 gpm. Also, the change in elevation is 495 – 365 = 130 ft.

Using the Hazen-Williams equation:

$$h_L = \left[4.73\,C^{-1.852}\,L\,D^{-4.87}\right]Q^{1.852}$$

$$= \left[4.73 \times 140^{-1.852} \times 15{,}000 \times \left(\frac{8}{12}\right)^{-4.87}\right]\left(\frac{500}{7.48 \times 60}\right)^{1.852}$$

$$= 4.73 \times 0.000106 \times 15{,}000 \times 7.03 \times 1.22$$

$$\simeq 65 \text{ ft}$$

The pressure at the end of the line will equal the change in elevation (130 ft) minus the head loss from pipe friction (65 ft) when converted to psi.

$$\text{Pressure} = 130 - 65$$

$$= 65 \text{ ft}$$

$$= \frac{65}{2.3} \text{ psi} \quad (1 \text{ psi} = 2.3 \text{ ft})$$

$$= 28.3 \text{ psi}$$

THE CORRECT ANSWER IS: (B)

502. Reference: Chen, *Civil Engineering Handbook*, p. 1129.

Find allowable head loss in line from Point A to warehouse:

$$h_{all} = 285 - 80.4 - (120 + 20 \times 2.31)$$
$$= 285 - 80.4 - 166.2$$
$$= 38.4 \text{ ft}$$

Using the Hazen-Williams equation:

$$h_L = \left[4.73\, C^{-1.852}\, L\, D^{-4.87}\right] Q^{1.852}$$

$$38.4 = 4.73 \times 130^{-1.852} \times 3{,}000 \times D^{-4.87} \times \left(\frac{3{,}500}{7.48 \times 60}\right)^{1.852}$$

$$38.4 = 4.73 \times 0.0001216 \times 3{,}000 \times D^{-4.87} \times 44.87$$

$$= 77.43 \times D^{-4.87}$$

$$D^{4.87} = \frac{77.43}{38.4} \quad \text{or} \quad D = 1.15 \text{ ft} \sim 13.8 \text{ in.}$$

THE CORRECT ANSWER IS: (B)

503. Reference: Merritt, *Standard Handbook for Civil Engineers*, p. 21.26.

Determine the combined minor loss coefficient for the gate valve, the globe valve, the pipe entrance and the pipe exit.

$K = (1.15 + 10 + 1.0 + 0.5) = 12.65$

$h_L = K V^2/2g$

where:

h_L = head loss

K = minor loss coefficient

V = velocity = Q/A

g = acceleration of gravity

Q = flow rate

A = area of flow

$h_L = 12.65 \times (Q/A)^2/(2 \times 32.2)$

$= 12.65 \times (7/3.14 \times 0.5^{2})^{2}/(2 \times 32.2)$

$= 12.65 \times (8.912)^2/(2 \times 32.2)$

$= 12.65 \times 79.43/(2 \times 32.2)$

$= 15.6$ ft

Find the head loss in the pipe using the Darcy-Weisbach equation:

$h_L = f LV^2/(2Dg)$

where:

h_L = head loss, ft

f = friction factor

V = velocity, ft/sec

g = acceleration of gravity

D = diameter, ft

$h_L = 0.015 \times 2{,}000 \times 79.516/(2 \times 1 \times 32.2)$

$= 37$ ft

Total head loss = 37 + 15.6 = 52.6 ft

The water elevation of the lower reservoir is equal to the water elevation of the upper reservoir minus the total head loss which is 100 ft – 52.6 ft = 47.4 ft.

THE CORRECT ANSWER IS: (A)

504. $V = \frac{1.486}{n} R^{2/3} S^{1/2}$

where:

$n = 0.012$ $S = 0.35\% = 0.0035$ ft/ft

R: Use Appendix A from Chow, *Open-Channel Hydraulics*: Geometric Elements for Circular Channel Sections.

for: $y/d_o = 3/15 = 0.2$ $d_o = 15$ in. = 1.25 ft

$R/d_o = 0.1206$

$R = 0.1206\,(1.25) = 0.151$

$$V = \frac{1.486}{0.012}(0.151)^{2/3}(0.0035)^{1/2}$$

$V = 2.08$ fps

THE CORRECT ANSWER IS: (B)

505. Reference: Mays, L. W., *Water Resources Engineering*, 1st ed., pp. 51, 88, 89, and 93.

Use Manning's equation to calculate the velocity of flow, and then calculate the Froude number.

Manning's equation:

$$V = \frac{1.486}{n} R^{2/3} S^{1/2}$$

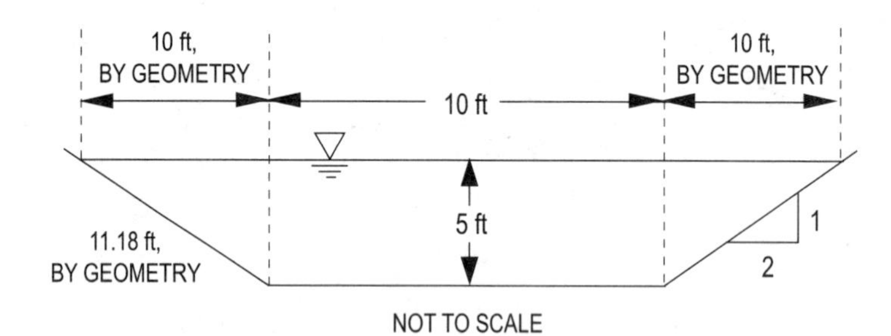

where:

V = velocity, fps

n = coefficient of roughness, unitless

R = hydraulic radius, ft = $\frac{\text{cross-sectional area of flow}}{\text{wetted perimeter}} = \frac{A}{P}$

S = slope, unitless

505. (Continued)

By geometry of the trapezoid:

Area = A = 1/2 (bottom width + top width) × height = 1/2 (10 + 30) × 5 = 100 ft^2

Wetted perimeter = P = 11.18 + 10 + 11.18 = 32.36 ft

R = A/P = 100/32.36 = 3.09 ft

$$V = \frac{1.486}{n} R^{2/3} S^{1/2} = \frac{1.486}{0.012} \times 3.09^{2/3} \times 0.001^{1/2} = 8.31 \text{ fps}$$

$$D_h = \text{hydraulic depth} = \frac{\text{cross-sectional area of flow}}{\text{top width of flow}} = \frac{100 \text{ ft}^2}{30 \text{ ft}} = 3.33 \text{ ft}$$

Froude number:

$$F_r = \frac{V}{\sqrt{gD_h}}$$

where:

V = velocity

g = gravitational constant

D_h = hydraulic depth = $\frac{\text{cross-sectional area of flow}}{\text{top width of flow}}$

$$F_r = \frac{V}{\sqrt{gD_h}} = \frac{8.31 \text{ fps}}{\sqrt{32.2 \text{ ft/sec}^2 \times 3.33 \text{ ft}}}$$

$$F_r = 0.80$$

THE CORRECT ANSWER IS: (D)

506. Reference: Chow, *Open-Channel Hydraulics*, pp. 75 and 362.

$Q = 3.0 \text{ MGD} = 4.62 \text{ cfs}$

$Q = 3.07 \times H_a^{1.53}$

$$H_a = \left(\frac{4.62}{3.07}\right)^{1/1.53} = 1.31 \text{ ft}$$

Elevation at A = 100 ft + H_a = 100 + 1.31 = 101.31 ft

$Q = 3.33 \times LH_b^{3/2}$

$$H_b = \left(\frac{4.62}{3.33} \times \frac{1}{4 \text{ ft}}\right)^{2/3} = 0.494 \text{ ft}$$

$$\frac{H_b}{H_a} = \frac{0.494}{1.31} = 0.37 \leq 0.6$$

Elevation at B = 100.2 ft + H_b = 100.2 + 0.494 = 100.69 ft

THE CORRECT ANSWER IS: (C)

507. Reference: Maidment, *Handbook of Hydrology*, p. 18.3.

To get a high degree of confidence (90%) of not exceeding culvert capacity in 30 years, you must select a much higher design return period than 30 years.

Math solution:

$$P = (1 - 1/T)^n$$

where:

P = probability that a storm will not be exceeded in n years
T = return period storm

$$0.9 = (1 - 1/T)^{30}$$

$$T = 285 \text{ years}$$

Another way to check is:

$$P = \Sigma X_p$$

where:

X_p = probability that the event occurs in any one year
T = 285

$$X_p = 1/T$$

$$P = (30)(1/285)$$

$$P = 0.10$$

Therefore, the probability that the event does not occur is $1 - P = 0.9$ or 90%.

THE CORRECT ANSWER IS: (D)

508. Reference: Maidment, *Handbook of Hydrology*, p. 9.30.

Q = effective rainfall × unit hydrograph ordinate (runoff/inch of rainfall)

Flow during the second hour: $Q = 1.5 \times 1.2 + 0.7 \times 0.5 = 1.8 + 0.35 = 2.15$ cfs

THE CORRECT ANSWER IS: (B)

509. Reference: Chen, *The Civil Engineering Handbook*, p. 1039.

Using the SCS method for an agricultural watershed, S = potential maximum retention after runoff begins (in.). Pick soil group D for swelling clay and pasture land in fair condition. Therefore, pick CN halfway between poor and good or 84.5.

$$CN = \frac{1{,}000}{(S+10)}$$

$$S = \left(\frac{1{,}000}{CN}\right) - 10 = \left(\frac{1{,}000}{84.5}\right) - 10 = 1.83$$

THE CORRECT ANSWER IS: (B)

510. Reference: Freeze and Cherry, *Groundwater*, 1979, p. 16.

Applying Darcy's equation:

$$V = -K\frac{dh}{dL} \approx -K\frac{\Delta h}{\Delta L}$$

$$= 15 \text{ ft/day}\left(\frac{200 \text{ ft}}{12 \times 5{,}280 \text{ ft}}\right)$$

$$= 0.047 \text{ ft/day}$$

$$\therefore \text{Travel time} = \frac{12 \times 5{,}280 \text{ ft}}{0.047 \text{ ft/day}} = 1{,}348{,}085 \text{ days} = 3{,}693 \text{ years}$$

THE CORRECT ANSWER IS: (D)

511. Reference: Metcalf and Eddy, *Wastewater Engineering*, 4th ed., p. 691.

Let design return activated sludge flow rate = Q MGD
Mass balance on solids: $8.34 \times 2{,}500 \times (3.5 + Q) = 8.34 \times 7{,}790 \times (Q + 0.07)$

Solving the above gives Q = 1.55 MGD

THE CORRECT ANSWER IS: (B)

512. According to the stoichiometric equation, 1.08 moles of methanol are consumed by 1 mole of NO^-_3. The molecular weight of methanol is 32 and that of nitrate-nitrogen is 14.

∴ Minimum amount of methanol required

$$= \left[\frac{1.08 \times 32}{14}\right] \times 25 \text{ mg/L} = 61.71 \text{ mg/L}$$

THE CORRECT ANSWER IS: (D)

513. Reference: Metcalf and Eddy, *Wastewater Engineering*, 3rd ed., pp. 409–410.

Q = 5 MGD/2 parallel trains = 2.5 MGD per train

$W = 8.34\ Q\ C = 8.34 \times 2.5\ \text{MGD} \times 200\ \text{mg/L} = 4{,}170\ \text{lb/d}$

F = 2 (given)

$$E_1 = \frac{100}{1+0.0561\sqrt{\frac{W}{VF}}}$$

$$70 = \frac{100}{1+0.0561\sqrt{\frac{4{,}170}{2V}}}$$

$$70+3.927\sqrt{\frac{4{,}170}{2V}} = 100$$

$$\sqrt{\frac{4{,}170}{2V}} = 7.6394$$

$$\frac{4{,}170}{2V} = 58.3607$$

$$2V = 71.4522$$

$$V = 35.7\ \text{thousand ft}^3$$

$$V = 36{,}000\ \text{ft}^3$$

THE CORRECT ANSWER IS: (C)

514. Reference: Viessman and Hammer, *Water Supply and Pollution Control,* 4th ed., p. 742.

Perform a mass balance at a point 30 miles downstream

Convert 100 cfs to MGD

$100\ ft^3/sec \times 7.49\ gal/ft^3 \times 1\ MG/10^6\ gal \times 86{,}400\ sec/day = 64.7\ MGD$

Calculate the mass of oxygen in the river just before the discharge from Industry Y

Total flow = river + Industry X

Total flow = 64.7 MGD + 5 MGD = 69.7 MGD

Mass of oxygen = 69.7 MGD × 4 mg/L × 8.34 lb/(MG•mg/L) = 2,325 lb/day

Calculate the required mass of oxygen

74.7 MGD × 4.25 mg/L × 8.34 lb/(MG•mg/L) = 2,648 lb/day

Calculate the oxygen mass difference

Difference = 2,648 – 2,325 = 323 lb/day

Calculate the amount of oxygen added from original discharge from Industry Y

Oxygen in discharge = 5 MGD × 3 mg/L × 8.34 lb/(MG•mg/L) = 125 lb/day

Calculate the mass of oxygen to be added

Oxygen to be added = 323 – 125 = 198 lb/day

THE CORRECT ANSWER IS: (B)

515. Phosphorus gets converted to chlorophyll (algae) until phosphorus concentration limits the process (Chen, *Civil Engineering Handbook*, p. 214).

After mixing:

Flow = 50 + 20 = 70 cfs

Mass balance on P:

$(50 \times 0.02) + (20 \times 0.5) = 70 \times P$

$$P = \frac{(50 \times 0.02) + (20 \times 0.5)}{70}$$

$$= 0.16 \text{ mg/L, which is greater than } 0.015 \text{ mg/L}$$

Therefore, algae will grow and consume the P; as the P drops below 0.015 algae will begin to die off.

THE CORRECT ANSWER IS: (D)

516. Convert MGD to cfs: 10 MGD ⇒ 15.45 cfs

In-stream temperature after mixing = [(15.45 × 42) + (100 × 19)]/(15.45 + 100)
= 22°C

Use $k_{1_T} = k_{1_{20}} \theta^{T-20}$ Eq. 2-25, Metcalf and Eddy, *Wastewater Engineering*, 4th ed., p. 86.

k_1 at $22°C = k_{1_{20}} \times 1.056^{(22-20)} = 0.11 \text{ day}^{-1}$

THE CORRECT ANSWER IS: (B)

517. Sludge volume entering tank = total flow rate × solids concentration × conversion factor
= 2.1×10^3 ft^3/day × 172 mg/L × 28.32 L/ft^3 × lb/4.54×10^3 mg
= 2.253×10^7 lb/day

Sediment volume allotted = tank volume × reserved %
= 1.41×10^{10} ft^3 × 22% = 3.102×10^9 ft^3

Weight of sediment in tank = sediment volume allotted × solids density
= 3.102×10^9 ft^3 × 80 lb/ft^3
= 2.482×10^{11} lb

Tank service life = sediment weight ÷ sludge weight entering tank/day
= 2.482×10^{11} lb ÷ 2.253×10^7 lb/day
= 1.102×10^4 days
= 1.102×10^4 days ÷ 365 days/year
= 0.302×10^2 years
= 30.2 years

THE CORRECT ANSWER IS: (B)

518. 99.99% removal corresponds to a log inactivation of 4.0. From the table, for 4.0 log inactivation, a temperature of 15°C and a pH of 7.5, the value of C•t = 4 (mg/L) • min.

C•t = 4@15°C, 7.5 pH

Q = 350 gpm = 46.6 ft^3/min

Calculate velocity in pipe to calculate residence (detention) time:

$$V = \frac{Q}{A} = \frac{46.6 \text{ ft}^3/\text{min}}{\frac{\pi}{4}\left(\frac{8}{12}\right)^2} = 132 \text{ fpm}$$

$$\text{Residence time} = \frac{1{,}800 \text{ ft}}{132 \text{ fpm}} = 13.54 \text{ min}$$

$$C = \frac{4 \text{ mg/L} \bullet \text{min}}{13.54 \text{ min}} = 0.295 \text{ mg/L}$$

THE CORRECT ANSWER IS: (A)

519. Reference: Davis and Cornwell, *Introduction to Environmental Engineering*, 4th ed., p. 243.

The lime needed to remove carbonate hardness is

lime = CO_2 + bicarbonate + magnesium to be removed + excess
lime = 50 + 120 + 0 + 20
lime = 190 mg/L as $CaCO_3$

THE CORRECT ANSWER IS: (C)

520. Reference: Newnan, *Engineering Economic Analysis*, 2nd ed., pp. 89–116.

$P_B = \$10{,}000{,}000 + \$200{,}000\ (P/A, \underset{11.47}{20}, 6\%)$

$= \$12.29$ million

$P_L = \$1{,}000{,}000 + \$600{,}000\ (P/A, \underset{23.12}{40}, 3\%)$ This is for semiannual payments for 20 years.

$= \$14.87$ million

$P_B - P_L = 12.29 - 14.87 = \$2.58$ million

P_B is better by almost \$2.6 million.

THE CORRECT ANSWER IS: (A)

Sample questions and solutions are also available for the following PE examinations:

Chemical
Electrical & Computer
Environmental
Mechanical
Structural I
Structural II

For more information about these and other Council study materials, visit our homepage at www.ncees.org or contact our Customer Service Department at 800-250-3196.